VOICES

from the

PANDEMIC

VOICES

from the

PANDEMIC

⸺◦◦◦⸺

Americans Tell Their Stories of
Crisis, Courage and Resilience

Eli Saslow

DOUBLEDAY
New York

Book design by Michael Collica
Jacket illustration © Nathan Wyburn
Jacket design by John Fontana

Library of Congress Cataloging-in-Publication Data
Names: Saslow, Eli, author.
Title: Voices from the pandemic : Americans tell their stories of crisis, courage and resilience / Eli Saslow.
Description: First edition. | New York : Doubleday, [2021]
Identifiers: LCCN 2021010073 (print) | LCCN 2021010074 (ebook) | ISBN 9780385547000 (hardcover) | ISBN 9780385547031 (ebook)
Subjects: LCSH: COVID-19 Pandemic, 2020– —United States—Anecdotes. | Americans—Interviews. | United States—Social conditions—21st century.
Classification: LCC RA644.C67 S277 2021 (print) | LCC RA644.C67 (ebook) | DDC 614.5/92414—dc23
LC record available at https://lccn.loc.gov/2021010073
LC ebook record available at https://lccn.loc.gov/2021010074

MANUFACTURED IN THE UNITED STATES OF AMERICA

1   3   5   7   9   10   8   6   4   2

First Edition

Contents

Author's Introduction

The COVID-19 pandemic was the rare international event that became personal to each of us. We all have our own story of this crisis. Everyone suffered in some way, and many did so alone. The virus isolated us in our own homes, our own bubbles, our own pods, our own personal hardships, our own ideological bunkers. The world contracted. The reporting in this book began as an attempt for me to see and feel beyond my own living room into the millions of personal pandemics unfolding all across America.

I spent hundreds of hours interviewing people from every corner of the country to collect the first-person accounts in this book. The words that follow are theirs. The stories are theirs. These are not retrospective accounts but interviews that occurred in the moment, as people encountered the disaster in real time. I talked to a small-town coroner as he moved the bodies of his friends and relatives; to a laid-off security guard as he waited in his first food line; to an anesthesiologist as he rushed between COVID intubations; to a grandmother as she was being evicted from her one-bedroom apartment; to a fifty-two-year-old as she spent her final days alone inside her hospital room. My interviews with each of the forty or so people included in this book typically stretched out over many hours and several days until I had dozens of pages of transcription. Then I condensed, structured, fact-checked, and lightly edited those transcripts into the monologues collected here.

Together these pieces tell the story of a country in crisis, but this book does not attempt to be a comprehensive history of the plague we're still living through. These are intimate stories from the country's hidden corners, about how the pandemic was personally endured and often overcome.

I worried when I first began this project that a year of so many intense conversations could become depressing, or even numbing, but in fact it was the opposite. Even in the midst of their own acute pain, the people in this book inspired me with their ability to trust, their empathy, their insight, their candor and emotional courage. In a miserable year made worse by so much political and systemic failure, it was their humanity that restored me. I hope it does the same for you.

VOICES

from the

PANDEMIC

Briefings from the World Health Organization

Jan. 4, 2020
"DISEASE OUTBREAK: pneumonia of unknown cause—China. There is limited information to determine the overall risk."

Jan. 30
"Patients often present with fever, cough and shortness of breath. The symptoms range from mild to severe disease, and some result in death. The cause of death is due to progressive respiratory and multi-organ failure."

Feb. 5
"There is evidence the virus is spreading by human-to-human transmission. The relatively small number of cases outside China gives us a window of opportunity to prevent this outbreak from becoming a broader global crisis."

Feb. 11
"We now have a name for the disease and it is Covid-19. We have a realistic chance of stopping this outbreak. Now is the time for solidarity."

Feb. 17

"The sudden increase in new cases around the world is very concerning."

Feb. 24

"Does this virus have pandemic potential? Absolutely. We must do more."

Feb. 27

"There are some vital questions that every country must be asking itself today: Are we ready for the first case? What will we do when the first case arrives? Do we have enough ventilators and other vital equipment? Do our health care workers have the training and equipment they need to stay safe? Do our people have the right information? Do they know what this disease looks like? Are we ready to fight rumors and misinformation with clear and simple messages that people can understand? Are we able to have our people on our side to fight this outbreak?

"These are the questions that will be the difference. If the answer to any of these questions is no, your country has a gap this virus will exploit. I repeat: If the answer to any of these questions is no, your country has a gap this virus will exploit."

Mar. 11

"We are deeply concerned both by the alarming levels of spread and severity, and by the alarming levels of inaction. We have therefore made the assessment that Covid-19 can be characterized as a pandemic.

"We cannot say this loudly enough, or clearly enough, or often enough: All countries can still change the course of this pandemic. We have rung the alarm bell loud and clear. This is not just a public health crisis. It is a crisis that will touch every sector and every individual."

Chapter 1

"Anything good I could say about this would be a lie"

Tony Sizemore, on the death of Birdie Shelton
Indianapolis—March 2020

She's dead, and I'm quarantined. That's how the story ends. I keep going back over it in loops, trying to find a way to sweeten it, but nothing changes the facts. I wasn't there with her at the end. I didn't get to say goodbye. I don't even know where her body is right now, or if the only thing that's left is her ashes.

From normal life to this hell in a week. That's how long it took. I'd barely even heard of this damn virus until a few days ago. How am I supposed to make any sense of that? It's loops and more loops.

She transported cars for a rental company. That's where all this must have come from. People fly in from somewhere for a meeting and fly out a few hours later. You've got germs from all over the world inside those cars. I didn't like the fact that she was working so hard, sixty-nine years old and still climbing in and out of Ford Fusions all day, driving from Indianapolis to St. Louis and back with bad knees, bad hips, diabetes, and all the rest of it. Sometimes she hurt so much after work that I had to help her out of the car. I guess I should have told her to quit, but nobody told Birdie anything. She liked to drive, and we needed the money.

I think she'd been feeling bad for a few days, but I don't remember much about what happened early on. She wasn't a complainer, and I'm not always the best at noticing. There was a cough somewhere in there. Probably a touch of a fever. But this was a few weeks back, when those things didn't mean so much. I thought she probably had a cold, or maybe bronchitis. She would get that sometimes, lose her voice and be fine a few days later, no big deal. But then she woke me up at about four in the morning and kept pointing to her throat. She said she couldn't sleep. Said her eyes hurt. Said it felt like somebody was pounding on top of her head. Birdie's usually one of those who wants to rub some dirt on it and keep moving, so when she told me to take her to the emergency room, I knew it was serious. I knew she was sick.

First it was a fever of 103. Then the doctors decided it was pneumonia and went ahead and admitted her. Then it was pneumonia in both of her lungs. If anybody was thinking it was the coronavirus, I didn't hear it—at least not at first. Nobody in Indiana had it yet. Nobody here was even talking about that. Maybe you'd see it on the news every once in a while. But even if it was killing a few people on some cruise ship or infecting a nursing home in Washington State, it was basically just happening on TV.

The best precautions weren't taken in the early stages.

Nobody really knew what the dangers were or what we were supposed to do. A few nurses wore gloves or masks when they came to see Birdie, but that seemed normal for treating pneumonia. I didn't wear anything to protect myself, and nobody really asked me to. I was lying next to her in the bed or sitting in a chair and holding her hand. She didn't have much other family, and if I got up to go out into the hallway for a few minutes, I'd lean down and kiss her goodbye.

Would it have gone any different if they knew what was making her so sick? Maybe. Or maybe they would have quarantined her right then, and I would have lost a few more days with her.

See, I could analyze this to death. I'll be doing this for the rest of my life.

It was hard for me to sit there. I'm almost ashamed to say that, but it's true. She was in the bed, and I was usually a few feet away in the recliner. It was two or three days in that room, but each one felt like a year. I'm not a natural caretaker, and never claimed to be, but it seemed like no matter what I tried, I couldn't help her. It was just watch, wait, touch her forehead, apologize, try to hold myself together. I couldn't do anything. Nobody could.

She was taking so much oxygen, but it was never enough. She had two little tubes put in her nose, and she still couldn't get enough air. They put a big mask on her face to get her oxygen back up, and that made her claustrophobic and panicky. She got real freaked out. I tried to count breaths with her to calm her down. I kept saying: "Easy. Easy. In, out. In, out." I couldn't distract her because she was so deep in her head with panic. It labored her to talk. It labored her to breathe. I said, "Don't talk then, honey. Just sit. Save your energy." There was a TV in there, but neither of us could focus on it. I sat in the quiet with her, for whatever comfort that might have brought her. I don't know. I listened to her breathing. I watched her. When she was asleep, she was taking these real quick, short breaths, like she was gulping air more than breathing it. When she was awake, she was kind of mumbling to herself. I tried to understand, but she wasn't making any sense. Maybe it was the medication they were giving her. I hope to God it was the medication. She was talking about how her eyes hurt, her insides hurt. She would clutch her fists and hit the bed and stuff, and you don't really know how to help somebody in that frame. I mean, when she's just clutching her fists and moaning and—I don't know. I don't know what I could have done. I sat there for as long as I could and then I got up every few hours to pace the hallway, or I'd drive eight minutes home to feed the dogs. I was starting to go a little crazy myself. I couldn't keep sitting there, feeling helpless, feeling useless, listening to her breathe.

It was an awful time. I should be thankful she's not suffering anymore, but she did suffer some.

It got worse. Her breaths got raspy. Shorter. They put her on life support. They rolled her across the hall one afternoon and tested her for the virus. At some point in there, I went home after midnight to check on the dogs, and when I came back early the next morning there was a sign that said "No Visitors" taped to the door of the hallway that led to her room. The whole thing became confusing to me. They said I couldn't go in. They said nobody could. I sat in the waiting room for hours. I peeked through the window down the hallway once and saw them moving her to a different room. It looked like she was sleeping with the tube down her throat. The doctor said she was heavily sedated to stay comfortable. I'd like to believe that, but I don't know if she was comfortable or not.

When they told me the COVID test was positive, I started learning more about the virus from the doctors. Right away I thought it was a death sentence for her. She had every underlying condition this virus attacks. Damaged lungs. High blood pressure. Her body wasn't strong enough. She was lying there waiting to die.

The doctors told me to go home, but I didn't. Most of the time, I sat by the elevators in the waiting room. Nobody else was in there. Sometimes one of Birdie's friends would come sit with me. The doctors kept saying "No change." "No change." It had been five days at that point since she'd woken me up pointing at her throat. They sent a chaplain to talk to me. He was wearing a mask, but he put his hand on my shoulder. Their voices kind of kept getting softer and softer. Everyone knew what was coming. I was up most nights and sleeping some in the day. My body wore down. I started coughing, and they told me I didn't have a choice. They said I needed to go home and quarantine. Here I was worrying that Birdie was going to be the first victim of this virus in Indiana, and now I'm thinking that I might be the second.

I walked circles in the house. I'm sixty-two, fairly healthy but not indestructible, and now I'm worrying about Birdie but also about my own mortality. Pretty soon the hospital started calling me to ask about unplugging her. They said her kidneys were shutting down. They said all of her organs were going, but that it was my decision. I told them: "How can I turn her off without looking at her? I can't take your word on this. She might be doing jumping jacks for all I know. I need to see her before I make that final call." That's when they started talking about setting up a video call, so they could take her off some of the meds and I could say some kind of goodbye.

I'm not a techie. Birdie liked to tease me about that. When I got together with her six years ago, I still had a flip phone. She liked going on Facebook, and to be honest it kind of pissed me off when she was on her phone all the time. The only friends I have are ones I can see and touch. But now I'm talking on the phone for hours with some nice lady from hospital IT, and she's telling me how to download some kind of app. She asked if I had an iPhone, but I don't. I found one of Birdie's old ones, but I didn't have the password. So now I'm getting frustrated, trying to get this video chat to work on my old Android so I can see Birdie on the ventilator. The lady said my phone needed to be charged to fifty percent for the video to work, but my phone hasn't seen fifty percent in two years. It only charges if it's turned off, so I started turning it off and back on every few hours to check if anyone had been calling about Birdie. And it's like, you know what, is this really what I should be doing right now? Is this really how I'm supposed to tell her goodbye? Over this damn phone screen? Finally the phone gets to fifty and then dies out of nowhere. Fifty again but I need some kind of password. I said to hell with it. This isn't going to work.

The doctor called the next morning. Birdie died at 10:20. So I didn't have to unplug her, and I didn't get to see her or say goodbye.

They held a press conference, since she was the first to die in Indiana. They told all the media that we got to say goodbye over video. I guess it's a nicer story. I don't really blame them. I'd like to find a way to sugarcoat this thing, too, but I can't. Anything good I could say about this would be a lie.

They told me to isolate and stay home for fourteen days from the last time I saw her. I'm kind of losing track of how many days it's been. I have some depression issues, and it would be real easy for me to go to bed and pull the covers up over my head. I could bury myself in this thing and let my mind keep running loops. I'm staring at her clothes in the closet. Her curling iron is on the bathroom sink. Her car's out front that we owe money on. I have no idea what I'm going to do. I don't know what bills are paid up and what ones aren't. She handled most of that. She looked after me in some ways like she cared for everybody, whether she knew you or not. That was always her nature. Anyway. I'll spend years trying to pick up all these pieces. Yesterday afternoon, they cut the power off, but I figured out how to get it back on.

I haven't eaten much, and it's probably making me weak. I'm bone-tired and coughing like crazy. They called me back to the hospital for a chest X-ray, but the doctors said I looked good. No fever. No trouble breathing. They decided not to even give me a test. They have twelve nurses quarantined over there now and a whole floor of people with the virus, but I got lucky. They told me I'll be fine.

Chapter 2

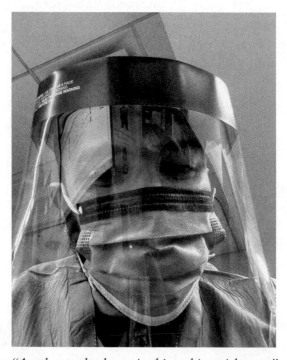

"Another twelve hours in this waking nightmare"
Sal Hadwan, on the failings of an overwhelmed hospital
Detroit—April 2020

I'm the shift leader of this nursing team. I give them their assignments at the beginning of the night, but lately there are nights when I can't find the words. I can't look them in the eye. There's so much shame because of the situation we've been put in. This virus is a monster to treat even if you've got every resource in the world at your disposal, and we're being sent out there half-naked. We're getting crushed every night.

I stand in front of the other nurses, stumbling through their assignments. "I'm sorry. I'm sorry you have to go out and deal with all this."

Overwhelmed would have been a good word for this hospital a week or so ago, when the pandemic was just starting. Now we're beyond that—we're way out of our depth. This virus went from being nothing but a rumor a month ago to taking over our whole emergency room. Sinai-Grace is the hardest-hit hospital in downtown Detroit, and it wasn't prepared. We're understaffed. We're running out of masks and protective gear. We have so many nurses out sick with the virus, and we haven't replaced any of them. Who wants to come work here right now? At best, it's going to be a nightmare. At worst, it's a death sentence. We don't have the resources to handle this. Each one of my nurses should be taking care of four to six patients during a shift. That's what it says in the handbook. I'm sending them out to care for fifteen or twenty people in critical condition. It's criminal. It makes me physically sick.

Some nights we've had 120 patients in the ER at once. That's three times what's normal. It's one emergency code after the next. We've been putting double beds in rooms, beds in hallways, beds in closets, until eventually we ran out of beds. Some people are sitting in wheelchairs in the hallway now. We have extension cords running everywhere to try to get oxygen to all these people. Patients sometimes sit in the waiting room for a full day after they've been admitted, because that's the only chair left. And these aren't normal patients. Half of them look like they're going to drop any minute. The acuity with this disease is way higher than what we normally see. It used to be that most people on ventilators would go up to the ICU, because that's their specialty, but now the ICU is full. We're basically handling the most severe cases in the ER, which is not our training. These nurses don't have a second to relax. You've got one patient's oxygen running out and another whose heart rate is going wild. All you can do is try your best to hear the alarms and then sprint as fast as you can from one emergency to the next. You hope you make it in time. Sometimes you don't.

Every single employee in this ER can tell a story of a patient that shouldn't have died on our shift. It's not these nurses' fault. They can't be fifteen places at the same time. I'm sorry, but they can't. It's not humanly possible.

The other night I bagged five bodies in the ER, and there was no place to put them. We have a morgue that fits twenty bodies, and that was full. So then we had to start putting them in the fridges outside in the parking lot, and that ran out of space. Then we started having to stack them up in the view-ing room next to the morgue, and then we took over the sleep study room, which isn't even cooled. Sometimes we've just had to leave them here in the ER, because we don't have the staff to move bodies. There's no dignity, and that's the saddest part. These people don't have any visitors and they are dying alone. We don't have time to sit and care for them and hold their hands. Some of these patients look terrified, and I don't blame them. That's what this virus looks like in here: oxygen masks and big, wide eyes. These patients see what's going on. I force myself to come back to work every night, and I wonder: Are any of the other nurses going to show up? Why would they? How could they?

And then they do show up, and they look at me, and I feel like I should have some kind of answers or a solution to give them. They deserve that. These nurses are so dedicated. They love this community, and that's the only reason they keep coming back. But none of the management in this hospital is here at night to deal with our concerns. Our ownership group is someplace out of Texas. We've complained and complained, but we've basically been left alone to cope with this. I'm just a nurse, and I have to be the one standing in front of them before we go back out onto the floor for another twelve hours in this waking nightmare. I'm not typically a guy for big speeches anyway, but there's not a motivational speaker in the world that could sugarcoat this. I just deliver it straight:

"Hey. I love you guys, and I'll have your back. I know we don't have time to take breaks, but try to remember to get a drink of water. You get one mask for your shift, so please don't lose it. Okay. I'm sorry. Let's go do the best we can."

MIKAELA SAKAL, nurse

This was my first nursing job. How crazy is that? I didn't sign up for this. Nobody signed up for this. Nothing that's happened here has gone by the book. Every night, we come into work and rewrite the rules.

The breaking point for me came last week. There were a lot of breaking points, but that was the last one. We got into work at seven, like always, and the first thing we do is get our assignment for the night. In school, what they teach you is it should be one nurse for every four patients in the ER. That's what you hope for. That's the ideal.

Our charge nurse, Sal, came in, and you could see he was upset. He told us: "These numbers are terrible. It's worse than ever. I don't even want to tell you how overmatched we are."

We had like seven or eight nurses staffing the entire ER. Some of us were going to have fifteen patients by ourselves, and that's when we decided: "Enough. We can't do this again.

It's not fair to us. It's not safe for the patients. We're not going
out onto the floor." Sal understood that we needed to take a
stand. We started calling and sending text messages to management: "We're not clocking in for work until you bring in
more staff." The day shift kept working overtime to support
us, which we knew was going to be brutal for them, but it
meant the patients were still getting care. All of us on the night
shift went into the break room. We listened to alarms going
off. We sat in our scrubs and we waited.

It's always been a little crazy working at Sinai-Grace, even
before all this. That's one of the reasons I came to work here.
They tell you: "This place will make you a great nurse." We
get more ambulances than any other hospital in Detroit. It's
sirens and resuscitations all night. I asked and I advocated for
myself to work in the most critical area, because I wanted to
learn. Nurses come here to get that hands-on experience, so
it was almost a point of pride sometimes if we were a little
short-staffed. Like, we can handle it. This is a tight-knit group.
We've been through a lot together. You think you've seen it
all, but then a week or so ago, the ER was suddenly getting
maxed out, and we had a bunch of staff leaving, or quarantined, or getting sick with this virus. Our patient loads started
going way up. Each night it was like: "Is it bad? Or is it really,
really bad?"

It got scary bad. I wish I could forget how bad it got.

Like the night it was just Joey and me assigned to take care
of twenty-six critical patients. Joey's one of our best nurses, and
I'd like to think we make a good team. We were in the part
of the emergency room called the TCU, or transitional care,
where they put the sickest people before transferring them to
the ICU. Usually, you might have ten patients in there, with a
few on ventilators who will transfer within a couple of hours.

This night we had eight on vents and the rest on supplemental oxygen. Some of the patients were awake, and some were sedated. A few had been in there for ninety hours. The ICU was full and we didn't have anywhere else to put people. There were stretchers lined up against the walls. We ran out of oxygen monitors. We were running low on oxygen.

You need to be everywhere at once. That's how it feels. You don't go to the bathroom. You don't eat. You spend every minute moving from patient to patient, trying to keep them alive.

There's constant noise, and it's all so mechanical. There's really no talking. The patients are using all their energy just trying to breathe. Most of them are too sick to ask for what they need. But call lights are going off and the alarm is beeping every time we get another medical trauma, which happens with this virus like fifteen or twenty times a night. The phones ring all the time, and it might be a family member asking for an update, but you look at the number and if it isn't a doctor, you honestly don't have time to pick it up. It's a family that might not get a chance to say goodbye, but you have to keep moving. Alarms are going off every minute. Pump alarms for the patients' life-sustaining medications. Monitor alarms. Oxygen alarms. Heart-rate alarms. Some beep, some chime, some ring. Every one could mean a crisis. I'd go home at night and hear phantom alarms. Sometimes I think I'm hearing them in my sleep.

And the thing is, you have to prioritize. You have to choose. You want to sit with these patients and build relationships and comfort them. That shouldn't be a luxury. That's part of basic care. What are you supposed to tell your patients when you might not have time to take them to the bathroom, or clean them, or call their families, or make them comfortable?

Some of these patients are hanging on and continuing to suffer because they don't have family with them. They need someone to say "It's okay. I'm here." They need someone to touch them. We had one nursing-home patient whose heart rate dropped really low, and he wasn't verbal at all, and you could see that he was scared and confused. His family wishes were that he didn't want to be intubated. We gave him a low dose of morphine for comfort. We stood in the hallway with him and took his hand and kind of rubbed his head, and as soon as we did that, this guy started to let go. We were able to be there for him, and a lot of times now, we can't be.

That's probably my best memory in all this.

There are a lot of bad ones. I had a patient in a back room, and her blood-pressure medication ran out. I was taking care of somebody else, because we're always taking care of someone else. We're changing an oxygen tank, or helping intubate someone, or refilling a crucial medication. We can't be everywhere. It's unrealistic and dangerous to keep this up. I heard the pump alarm in her room at the last minute. By the time I got there, her pressure had dropped to like 40 over 20. She was still alive, but barely, and I don't know how she's doing now. She might never recover from a crash like that.

There was another patient a few rooms over, and he was in bad shape. Joey had to leave the floor to transport someone to the ICU, so now I was alone with twenty-five or twenty-six people. It was maybe five o'clock in the morning. At that point, I'm ten hours in, and I'm exhausted. I was responding to alarms and trying to keep an oxygen mask on one lady who was confused and kept wanting to take it off, even though her life depended on it, and meanwhile, this other patient was in a room pretty far out of sight because we didn't have any

information about her condition. I called nurses and doctors. I went in wearing my uniform and tried to act like I was there for something official, but they weren't allowing visitors, so that didn't go too far. I have a cousin who plays in the NBA, and we had his team doctor calling the hospital for answers. You'd get somebody to pick up every five or six times, and we got information in snippets. First they said she had pneumonia, and it was bad. Then they said she needed to be on a ventilator, and we agreed to that. Then they told us she had about a ten percent chance to live. I'm like: "Really? She's sixty-eight. She's healthy. What happened?" We had no way to talk to her or see her, so we had nothing else to go on. It was breaking my family apart.

I went up to the hospital one day and tried to talk my way in, but they wouldn't budge. I said: "Are you really telling me my mother is going to die alone without me saying any kind of goodbye?" I went out to the parking lot and sat in my car. I'm someone who used to see my mom every few days, you know? I prayed. I cried and I meditated. I sat there in the driver's seat and had a long conversation with her. I know she couldn't hear me, but I needed to be close. I told her: "If you're holding on for us, you can let go. If you're suffering, let go."

She died a few days later. That's when the nightmare really got started.

We wanted to have a viewing and a nice service, and the hospital told us we could pick up her body after forty-eight hours. They said they needed to put her on ice to let the virus decontaminate. I don't know the medical stuff, but that's what they said. So we made all of the arrangements with the funeral home, and after two days they drove over to the hospital to pick her up. They waited for a while, and then the hospital told them her body wasn't ready. They said to come back in a few hours. So, fine. They came back, and the same thing happened. This time the hospital said it would be another forty-eight hours before we could get her, so now I'm getting suspicious.

I kept calling the hospital. Nobody even bothered to pick up. Another day came and went, and the funeral home said they still didn't have her. Then it was another day. Then another. At this point, it had been almost a week since she died, and we were putting these funeral arrangements on hold, and I was livid. What is happening here? Where is her body? My pops is a pastor. The funeral, the afterlife—this stuff is important to him. He's a calm person, but I don't stop to think when I get mad. I don't rationalize. My anger is going to be unleashed on the first person that I feel like is a part of it. I went up to the funeral home screaming and yelling, and finally they gave it to me straight. They said: "Her remains have been lost. The hospital can't find her remains."

What do you say to that? The confusion, the rage—my face gets hot just thinking about it.

I called the head of security at the hospital. I was losing it. I called him every name in the book. He said they would look for her body right away. I was like: "What the hell do you mean you'll look? How can you lose a person? How can you lose her?" I said I would give them an hour. After that, I was going to start calling lawyers. All hell was about to break loose.

I got a call back within about forty-five minutes. They said they'd found her remains and they were sorry for the delay. They didn't tell me anything else. I saw pictures on the news of bodies stuffed everywhere in that hospital. I don't know if she was in a closet or a storage room or what. To be honest, I don't want to know. I'm not sure I can handle it. The security guy just told me, "Okay. We've located her remains," and that was it.

Remains. They kept using that word like it might make it less personal. But these were bodies stacking up. These were people.

Chapter 3

"How are we supposed to eat?"

Burnell Cotlon, on how his grocery
store became a food pantry
New Orleans—April 2020

I know every person who comes into this store, and they know me. Burnell's Market. It's my name above the front door. These are my neighbors, but now we're eyeing each other like strangers, paranoid and suspicious. "Don't stand so close." "Don't breathe too heavy." "Just drop the groceries in the trunk and walk away." Some people have started sliding money back and forth to me across the counter with a plastic spoon.

Everybody's scared of everybody in a grocery now. There's so much fear, and I get it. I'm scared of catching this virus, too. But what bothers me more is the desperation.

This is one of the only fresh groceries in the Lower Ninth Ward of New Orleans, so pretty much everybody's a regular. They come for cigarettes or a biscuit on their way to work. It's mostly tourism jobs down here—Bourbon Street, the big hotels, the seafood industry and restaurants downtown. Those places have all closed up in the last month. At least half my customers have lost their jobs. They come to the register counting food stamps, quarters and dimes. A lot of people cry at my register and start apologizing. I keep telling them, "It's okay. I'm not in a hurry. Take your time. Stop apologizing." I had somebody barter me last week over a $0.75 can of beans. I used to sell two pieces of fried chicken for $1.25, and I cut it to a dollar.

We have an ATM in the store, and I watch people punching in their numbers, cursing the machine, trying again and again. It gives out more rejection slips than dollar bills. A lot of these people don't have any savings, and what they had was gone within that first few weeks after the city shut down. There's no fallback.

Last week, I caught a lady in the back of the store stuffing things into her purse. We don't really have shoplifters here. This whole store is two aisles. I can see everything from my seat up front. So I walked over to her real calm and put my hand on her shoulder. I took her purse and opened it up. Inside she had a carton of eggs, a six-pack of wieners, and two or three candy bars. She started crying. She said she had three kids, and her man had lost his job, and they had nothing to eat and no place to go. Maybe it was a lie. I don't know. But who's making up stories for seven or eight dollars of groceries? She was telling me, "Please, please, I'm begging you. How are we supposed to eat?" I stood there for a minute and thought about it, and what am I going to do?

I said: "That's okay. You're all right." I let her take it. I like to help. I always want to say yes. But I'm starting to get more desperate myself, so it's getting harder.

The first time in my life I let a customer float on credit was four weeks ago. It was a young guy who comes by most days to buy a few things, or just to sit outside with me on the milk crates and chop it up. He got home from the military a few years back, and I was in the Army, so we have that in common. Good guy. He'd been working as a cook downtown, and last month his restaurant closed up when the virus started to get real bad. He asked if he could come work for me.

I operate on a shoestring. It took every dollar I had to open this place after Hurricane Katrina. I had eighty thousand in savings after I got out of the Army, and all of it went into this place. I've robbed Peter to pay Paul so many times that Peter's got nothing left. So I had to tell him: "I already have a cook, and I'm barely paying him. I can't afford to pay two." But this guy, he was hurting. He needed something to eat. He picked up four cans of tuna, a Sno-Ball, and laundry detergent. He told me he was good for it as soon he gets his first unemployment check, and I trust him. I rang it up for $11. I took out a notebook that I usually keep near the register and started a little tab.

That notebook kept coming back out. Next it was Ms. Richmond. She did housekeeping at a hotel and lost that. Her tab was $48. Then it was a lady who shucks oysters downtown. She's got a big family to take care of, so she's at $155. Then there's another guy who I deliver to, since he's bedridden, and I showed up with two bags and he had nothing to give me. So he's at $54.80.

This has gone from a grocery store to a food pantry. That's how I'm feeling.

And what am I supposed to say? I don't blame any of these people. I feel for them. They're my friends and my neighbors. Some of these customers, I love. I truly do. They're getting by however they can. It's not their fault that just about every restaurant job in this city has disappeared. It's not their fault they still have to eat. They're not coming in and asking me for handouts on gin or beer. I don't sell alcohol. I won't give loans on cigarettes. What they need is milk, cheese, canned goods, bread, toilet paper, bleach, baby wipes. It's basics—the basic essentials. One elderly guy tried to start a tab for $3 of snacks for his grandchildren, so I went ahead and gave him the snacks. Another lady said she lost her job at a nightclub downtown, and she tried to proposition me for $20 even though my wife was standing right there working at the register. I gave her the $20 and she left.

I've got sixty-two tabs in the book now. From zero to sixty-two in less than a month. It's page after page of customers on credit. I'm out almost $3,000 so far. I know that might not sound like much, but at a little corner store like this it's my electric bill, my water bill, the mortgage on my house. I never missed a mortgage payment in my life until April 1 just came and went, and now this virus has me calling around and asking for forgiveness, too. The power is out now in my little laundromat and I can't afford to turn it back on. I'm paying one of my employees with free breakfast. I'm maxed on bills. I'm doing my best to keep this place open. Everybody here is waiting on unemployment checks and stimulus payments to keep us going, but let's be real. Some of these losses aren't ever coming back. I know how this goes. I lived in a FEMA trailer for three years after Katrina. I went from having forty-eight neighbors on my block to having three. They can talk all they want about how we'll bounce back and this will all be behind us before we know it, but not everybody bounces back. Some people are already standing in quicksand. There might be a recovery on Bourbon Street, but when will it show up here? Recovering can take forever.

Sorry. I try not to be angry. There's no use in it, and it's not my personality. I'm an optimistic person. Before this, I was always real upbeat. My wife tells me I'm optimistic to a fault.

She thought I was insane to open this store. Everyone did. When I moved back to the Ninth Ward after Katrina, there was nothing here. It was worse than starting over. This whole neighborhood was just rotted-out buildings and overturned cars. I lost my house. I lost everything except my wallet and three pairs of jeans. There were only a few hundred people who came back to live over here, because you had to fight. You had to really want it. If you needed to go to work, you had to walk a mile to the bus stop and then hope the bus was maybe going to show up. If you wanted some kind of snack or fresh produce, you had to get on the bus and then transfer three times to get to the Walmart in Chalmette. This whole part of town was a food desert. People weren't coming back because we didn't have any stores, and the chain stores weren't coming back because they said there weren't enough customers to justify it. So what comes first? There was no chicken and no egg. I kept calling Walmart and Winn-Dixie and begging them to build a store. I got the mayor involved. It's not like I began this whole thing wanting to become a grocer. I've got my degree in criminal justice. But food is the number one essential. If you don't have a place to get yourself something to eat, you don't have a neighborhood. Somebody had to do something.

I bought an old building eight blocks from the levee. It wasn't up to code. It wasn't up to anything. It was a burned-out pool hall without half a roof, and it took me four years to get it functional. I didn't know what I was doing. I barely knew what a hammer was. I took supplies out of houses the city was bulldozing and then tried building stuff myself. I built my own shelves, and they turned out kind of lopsided. I went on YouTube and worked out how to do wiring, because we had no electricity. My floor is uneven because I didn't know how to pour the concrete. It was trial and a whole lot of error, and this place still doesn't look like much. To a lot of people it's just plain ugly, but I say it's got character. It was the best day of my life when I finally opened it up.

My customers come to the store because it's a happy place, and this community deserves something good. I'm in here seven days a week, from sunrise to about ten at night. The hours are pretty flexible. The store closes whenever I go home, so I try to stay as long as I can. I have a lot of people who come in here three or four times in a day, because there's still no place else to go, and they want that chance of having a conversation. This is their store. That's how it feels to people in this neighborhood. I've got music blaring out front, always upbeat, drawing people in. I have candy and cold treats for the kids that are out of school. I'm running out of some things now that it's getting so tight. I'm low on rice and sugar, but I hustle to fill this store. I say to my customers: "Tell me what you want and I'll stock it." They're grateful for this place. But right now the whole mood has changed. My number one seller is the two-dollar masks that I keep on the counter by the register. In some ways, this whole pandemic is harder than Katrina, because you can't evacuate. The virus is invisible, and you can't see it coming. Most everybody coming into the store is terrified. They're upset. They're mad. The reality is, this virus is hitting the Black community harder. It's the same old story.

Life in this neighborhood is an underlying condition: hard jobs, long hours, bad pay, no health insurance, no money, bad diet. That's every day. They have disabilities. They have high blood pressure, breathing problems, diabetes. I gave free blood-pressure checks when I opened, and not because it was good for business. Cigarettes are a big seller. Candy and cold drinks go quick. I tell people, "You only live one life. You've got to look after it." But fruits and veggies are expensive. If you're hungry, are you spending that dollar on an onion, or on nachos with chili cheese? We were made more vulnerable to this virus down here because of what we've had to deal with. Wearing a mask won't protect us from our history.

All of us know people now who are sick, or worse. My mom was exposed and quarantined for two weeks. I had a guy in the store the other day talking about how his sister was going on a ventilator. We lost one of our customers a few days ago, Mr. Lewis. He ran a free museum on Black culture. Sixty-eight years old, and that's that.

There's another lady who lived two blocks from the store. She'd been coming almost every week since I opened, but lately she'd been having a hard time. She lost her income and needed groceries, so we started her on a tab. Then she caught the virus, and I delivered more groceries to her porch.

She died last week, and a few days later I went into the book to look at her tab. There are a few accounts closing like that now, and probably more coming. Hers was 72 dollars and 14 cents. I found her name and drew a line through it.

Chapter 4

"You're basically right next to the nuclear reactor"

Dr. Cory Deburghraeve, on the
pandemic's most dangerous job
Chicago—April 2020

I could be the last person some of these patients ever see, or
the last voice they hear. A lot of people will never come off
the ventilator. That's the reality of this virus. I force myself to
think about that for a few seconds each time I walk into the
ICU to do an intubation.

This is my entire job now. Airways. Coronavirus airways.

I'm working fourteen hours a night and six nights a week. When patients aren't getting enough oxygen, I place a tube down their airway so we can put them on a vent. It buys their body time to fight the virus. It's also probably the most dangerous procedure a doctor can do when it comes to personal exposure. I'm getting within a few inches of the patient's face. I'm leaning in toward the mouth, placing my fingers on the gums, opening up the airway. All it takes is a cough. A gag. If anything goes badly, you can have a room full of virus.

So, there's a possibility I get sick. Maybe a probability. I don't know. I have my own underlying condition when it comes to this virus, but I try not to dwell on that.

Up until a few weeks ago, I was the anesthesiologist people would see when they were having babies. I'd do five to seven deliveries a day, mostly C-sections and epidurals. We're a large state hospital at University of Illinois–Chicago, and we end up doing a bunch of high-risk deliveries. You're trained to be the calmest person in the room. They teach us: "Don't just rely on medication to calm a person. Use your voice, your eye contact, your whole demeanor." We give people positive ideas and positive expectations. It sounds kind of corny, but it works.

Our team had a meeting in the second half of March to figure out our staffing plan, once it was clear where this pandemic was going. Chicago's becoming a hot spot now. Our ICU is almost full with COVID patients. The pediatric ICU has been cleared out to handle overflow. The wave is starting, and we need to limit our exposure or we're going to run out of staff. Everyone basically agreed we should dedicate one person to COVID intubations during the day and another at night, and I started thinking: I'm thirty-three years old. I don't have any kids at home. I don't live with older relatives. About an hour after the meeting, I emailed my supervisor. "I'm happy to do this. It should be me."

Now my pager goes off throughout the night. Nine o'clock, midnight, two, then again at three-thirty. Most of the time, I do several airways in a shift. By next week or the week after that, they're saying it could be as many as ten.

It's a common procedure. Intubations are routine for us, at least most of the time. You can be in and out of the airway in ten or fifteen seconds if everything goes right. But when you're dealing with a patient who isn't getting enough oxygen—which is everyone at this point—every second becomes crucial. As soon as I get the page in my call room, I grab my backpack of medications and my duffel bag of protective gear and run for the stairs. There isn't time to wait for the elevator. I go two floors up to the ICU and get into my protective gear outside the room: mask, face shield, hood, secondary hood, personal air filter, gown, two sets of sanitized gloves. I'm lucky that I still have good protective gear, because other hospitals are starting to run out. I tape everything that I'm wearing together, because a few times the gown has risen up and exposed my wrists. There are so many opportunities to contaminate yourself. I monitor my heart rate, and it goes from like 58 to 130 by the time I get into the ICU. I'm stressed and rushed and hot inside the protective gear. I'm trying not to show it.

I've been shocked sometimes when I walk in and see the patients. Most of the ones I've intubated are young—thirties, forties, fifties. These are people who walked into the ER because they were coughing a day or two ago, or sometimes hours ago. By the time I come into the room, they are in severe respiratory distress. Their oxygen level might be 70 or

80 percent instead of 100, which is alarming. They are taking 40 breaths a minute when they should be taking 12 or 14. They have no oxygen reserves. They are pale and exhausted. The fatigue puts them in a mental fog, and sometimes they don't hear me when I introduce myself. Some are panicky and gasping. Others are mumbling or incoherent. Last week, one patient was crying and asking to use my phone so they could call family and say goodbye, but their oxygen levels were dropping, and we didn't have time, and I couldn't risk bringing my phone in and contaminating it with virus, and the whole thing was impossible. What was I supposed to do? I kept apologizing. I just—I don't know. I have to find a way to hold it together in order to do this job. I tear up sometimes, and if I do, it can fog up my face shield.

The first thing I do is pull up a stool and get right down to their level at the bed. Most of the time, the look in their eyes is fear. But sometimes, honestly, it is relief, like, "Thank God. I can't do this anymore." Some of these people are fighting for so long they have nothing left. There's a sense of surrender. They don't have the energy to be hysterical.

I put an oxygen mask on the patient and give 100 percent oxygen for a few minutes. You want to tank them up, because they won't be able to breathe on their own. Next I give medication to put them to sleep. We're trained to touch the eyelashes a bit to make sure they're down. Then I give a muscle relaxer and take a look down the airway for the vocal cords. With this virus, I see significant upper airway swelling, tongue swelling, lots of secretion. When I start to put the tube in, that gives an opportunity for the virus to release into the air. The patient's airway is wide open at that point—no mask or anything. People can cough when the tube goes in toward the trachea, a deep, forceful cough. My mask and hood can get covered in fluid. Usually it's tiny droplets. Aerosolized virus can float around. You're basically right next to the nuclear reactor. I go in confident and fast, because if you miss on the first try, you have to do it again, and then you're bringing out a ton more virus.

Once I'm done, sometimes I'll go back to the call room and do squats or lunges. I try to keep my lungs strong. It's hard not to think about, because I've had bad asthma since I was a kid.

I use an inhaler twice a day. I'm very in tune with my breathing, and whenever I'm getting sick, the first symptom is I start wheezing. My whole family was like, "Why are you volunteering for this? What are you doing?" My dad and brother got a bunch of tools and built a Plexiglas intubation box based on a model out of Taiwan. It sits above the patient's face, like a shield to reduce your exposure. I haven't been able to use it yet, but they're worried. They're trying to protect me.

Last week, I called to tell them about my end-of-life wishes. Then I emailed it all to them, just in case. I said, "If I have to be intubated, I'm fine with that. But if I'm going through liver and kidney failure, and if I'm cognitively impaired at that point, and if you can tell my body is failing and I'm not going to get back to being who I am . . ." Well. It was a hard conversation. But it's better to be prepared. I know how this virus can go.

Each night, I try to do rounds with the doctors in the ICU to check on the patients I've intubated. They're not allowed to have family or visitors. It's a pretty lonely place right now— just doctors and nurses in protective gear and a lot of beeps. I feel bad for the patients, even if they're medicated. I'm not a religious person, but I do like to stand there for a minute outside the room and think about them and what they're going through. I try to think about something positive—a positive expectation. Mostly they're unconscious on the vent, but each day for an hour or two, they get what we call a sedation holiday, which means we bring down their medications so we can check on their baseline level of consciousness and see how they're doing on their own. In other words, for a little while, they might wake up.

Some people fight the tube and try to take it out, so you might have to restrain them. Other people just kind of look around and space out. They can't talk with the tube in, but I have seen a few patients before write messages on a piece of paper. "Vent?" Or: "Surgery?" Or: "How much longer?"

Usually, before this, patients would be on a vent for three to five days. That's what we considered to be normal. Now we're seeing fourteen to twenty-one days, and sometimes it has gone for a month or more. Most of these people have acute respiratory distress syndrome. There's inflammation, scar tissue, and fluid building up in the lungs, so oxygen can't diffuse easily. No matter how much oxygen you give them, it can't get through. It's never enough. Organs are very sensitive to low oxygen. First comes kidney failure, then liver failure, and then brain tissue becomes compromised. Immune systems stop working. There's a look most people get, called mottling, where the skin turns red and patchy when you only have a few hours left. We have a few at that point in the hospital now. Some of them have been converted to "do not resuscitate."

In between intubations, I'll sit in my call room and watch the monitors. I can see all of the patients' vitals and check on how they're doing. We've had some successes. A younger patient came off the vent earlier this week and just got sent home, and the whole staff at the hospital clapped as he walked out. The staff at this hospital is amazing. Even so, it usually goes the other way with this virus. I'm looking at the monitor right now, and there's one patient who isn't going to make it through the night. Three others are tipping toward the edge.

It's a powerless feeling, watching someone die. The oxygen level drops, the heart rate drops, the blood pressure drops. These patients are dying on the ventilator, and sometimes when they take away the body, the tube is still in the airway.

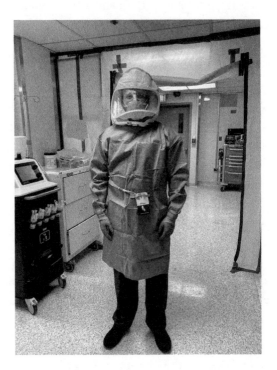

MOLLY DAWSON, patient from
Bowling Green, Ky., on being intubated

My oxygen level kept dropping, and I knew what was coming next. That tube was my biggest fear. I couldn't talk anymore because of the virus, but my thinking at that point was still

pretty clear. The doctor came in to tell me about the intubation procedure, and I wrote down a message on my phone and
showed it to him: "No tube. Please, please, no tube. Anything else."

I'm an ICU nurse at a hospital here. I know exactly what it
means to be intubated and put on a ventilator. I care for those
patients at the hospital every day, and I love what I do. But I
also see exactly what happens when you go on the vent. Over
the course of a few seconds, you go from being an entirely
independent person to being totally dependent. You're medically paralyzed. You're not breathing on your own. Your arms
are restrained to the bed, you're on all kinds of medications
and you don't know what's happening. You're stripped naked
and catheterized. You get a central line put in your neck, and
you're bathed to prevent all these possible infections. Every
sense of control you have is taken away, and you have no idea
when or if that will ever change. I've had a lot of patients who
don't come off the vent. You're giving up all autonomy. You
become part of a machine.

It's a last resort, but I got to that point really fast. I caught
the virus from a patient at work, and a week later I was spiking a fever of 103.5. I took a COVID test, and it was negative.
I took another—same result. It was driving me crazy, because
I knew something was very wrong. I'm only twenty-five. I'm
someone who likes to take charge and stay busy, so why am I
getting winded and holding on to the wall just to get myself
to the bathroom? I tried doing breathing treatments and managing it at home for a few days, but then I woke up one night
and I had this feeling like I was drowning. It was like breathing through the world's smallest straw. I drove myself to the
ER and sat in the car for forty-five minutes, because I didn't
have enough energy to walk inside. I was going to the same

hospital where I work, so everybody there knows me, and they looked at me and knew right away there was something seriously wrong. I was gasping. I was getting kind of purple. They did another COVID test, and this time it was positive. They took my pulse-ox and it was 78 percent, which is insanely low. If I saw that in one of my patients, I know what I'd be recommending. I started worrying about the tube.

They tried giving me an oxygen mask for a while, but that only got me up to 87, 88 percent. The air couldn't get through my lungs. They were too swollen from the virus. I kept watching the numbers on my screen. I knew in my gut what needed to happen, but I was trying to fool myself. I told the doctor to keep holding off. I kept writing him those little notes on my phone. I sent him out of the room at lunchtime and again in the early afternoon. I kept lying on my stomach to open up my lungs and get more air, but nothing really helped. I have seen a lot of patients with this illness who can compensate for a while and then they fall off a cliff, and I knew that cliff was coming for me. It felt like I was getting close to blacking out. My best friend is also a nurse at the hospital, and she came into my room and closed the door. She said: "I know you're scared, but this has to happen. We need to intubate now."

I started to cry from the anxiety. I couldn't really talk, but I nodded my head.

It takes fifteen or twenty minutes to set up an intubation. When you're the nurse helping on the procedure, that time goes really fast. You're running around to set up the room and get the medications. But when I was lying there, it seemed like forever. They moved me into a negative air pressure room. They found a Catholic priest and we did a video prayer. I texted my mom and told her I loved her. The doctor showed me the tube and talked me through what was going to happen, but I already knew, and I couldn't really focus on listening. I held my friend's hand before they gave me the medication. I told her: "I want to wake up someday, but don't let me wake up until it's over. Knock me out good. I don't want to feel the tube. I don't want to remember any of this."

They must have given the medication at that point, because everything went black. Or mostly black. The whole thing is a haze. I remember coming to at one point and feeling the tube in my mouth. I could tell that I was paralyzed from the medication, and I was in a different room. I didn't recognize any of the nurses around me. I was trying to talk, but I was suffocating, and it was hell, and then it was back to darkness again. I had a few dreams that I can remember. In one, I had lived my life and I was ninety years old and people were gathering for a party. In another, it was my funeral and they were playing this loud Dropkick Murphys song.

But most of what I remember is the nightmares and hallucinations. Those were the worst times of my life. I was lying on top of dead bodies and getting eaten by wild animals. I had trashed a hotel room. Someone was forcing me to take heroin. I was being chased by the police. It was all so vivid that I couldn't tell what was a dream and what was real. I think some of it was because I was coming down off the medications. The nurses told me that whenever they lowered my sedation, I would become agitated and start biting on the tube. I tried to yell. I tried to rip it out, and the doctors said eventually I got ahold of it and basically extubated myself. They decided to see how I did without the ventilator, and luckily I was starting to breathe a little better on my own.

When the tube was out and I could talk again, I asked a nurse how long I'd been under. I was thinking it had been hours, maybe a few days. They told me it had been almost three weeks. They said I'd almost died a few times, and my organs had started to fail. I'd been helicoptered to a different hospital. I didn't remember any of it.

It's been a month now, and my throat is still sore. My voice sounds scratchy. I can feel the air sliding up and down my throat when I breathe, and it's weird to think how much I took that for granted. I try to pay attention to it now. I'm still trying to reclaim some of my independence. I'm living with my parents and I'm doing physical therapy and occupational therapy a bunch of times each week. I'm getting to the point where I can go to the bathroom or eat by myself. It's little victories, but the doctors keep telling me to celebrate.

I don't know if it's PTSD or just part of recovering, but I have these big emotional swings. One moment, I'm like: "Wow. I'm so grateful. The intubation saved my life." But a lot of times I still have these nightmares, where the tube is still stuck somewhere in there, and I keep reaching in to pull it out.

Chapter 5

"I apologize to God for feeling this way"

Gloria Jackson, on being seventy-five,
alone, and thought of as expendable
Lake City, Minn.—April 2020

I try to remember that I'm one of the lucky ones in all this. What do I have to complain about? I'm not dead. I'm not sick. I haven't lost my job or gone broke. I'm bored and I'm lonely, and so what? Who's really going to care about my old-lady problems? Lately, when I see people talking about the elderly, it's mostly about how many of us are dying off and how we're forcing them to shut down the whole economy.

I tell myself I should be more positive. I should be grateful. Sometimes I can make that last for an hour or two.

A day can drag on forever when you're isolated all by yourself. I sleep as late as I can. I try not to look at the clock. I go on Facebook and read about all the ways this country is going to hell in a handbasket. I turn on the TV to hear a bit of talking. It's been almost seven weeks since I've spent time with a real, live person. I haven't touched or really even looked at anyone in what feels like forever, and it's making me start to think recklessly. The other day I went to Walgreens to pick up my medications, and I sat in the parking lot and thought about going inside. I was wearing my mask and I had my inhaler. I wanted to run a normal errand, look at the chocolates, maybe find my way into a conversation. But I stayed in the car and went to the drive-through. I put on my gloves and handed my card to the clerk through a hole in the glass window. I took the medicines and gave a little wave.

If I get this virus, I'm afraid it would be the end of me. I'm seventy-five. I've got all I can handle already with my asthma, fibromyalgia and an autoimmune disorder. The best way for me to survive is by sitting in my house for however many weeks or months it's going to take. It seems like you can catch this virus from anywhere, and it doesn't take much. I'm trying to make sure I don't come into contact with anybody. But how many computer games can you play before you start to lose it? How many mysteries can you read? I realize time is supposed to be precious, especially since mine is short, but right now I'm trying every trick I know to waste time away.

Negative thoughts creep up like that. I start getting crabby. It's waves of anger and depression, and I beat myself up for it. People have it a whole lot worse. Everywhere you look, somebody is really suffering right now. I know that. Obviously.

I've got two daughters out of town who call me and check in every week or so, but I don't want to guilt them. I've got a high school friend who dropped off groceries. I've got a dog and two cats that need to be cared for, which gives me some-thing to do. I've got my own manufactured home with flow-ers blooming all over the house. A lot of people don't realize there's a big difference between a trailer park and a mobile home community. I've spent hours lately driving up and down every block of this neighborhood, looking at people's yards, checking out whatever might be poking through the dirt. It's nice to see those little signs of life. One morning I drove my dog to the river thinking that maybe we'd get out and go for a little stroll. But other people were walking on the path, and I was worried about the viral droplets and all that. It didn't seem worth it. We sat in the car and cracked the windows and listened to the water.

It feels like everybody here is trying so hard to be cheerful, but boy does it take an effort. Our stores are closed. Our res-taurants are going out of business. The whole country is shut-ting down, and it's like we're saying goodbye to our old way of life. The other day was supposed to be the beginning of base-ball season, and I love baseball and I always look forward to that, and the anchor came on to the local news and said: "Let's all try to look on the bright side! Let's find a way to celebrate Opening Day even though nobody is playing." He showed pictures of fans wearing their Minnesota Twins T-shirts, or rubbing hand sanitizer onto a baseball to play catch, or grilling up some ballpark franks for dinner, and I thought: You know what I'd really like to do right now if I'm being honest? I'd like to find a bat and a ball and go break a few windows.

I apologize to God for feeling this way, but he made me how I am. I'm over this whole thing. Sometimes I start think-ing: What's the point of living if you can't see or touch any-body. Does that even count as being alive?

I used to be an optimist, but I'm not anymore.

I've never been this angry, and it's an ugly way to feel. Maybe when you don't get to see anybody for weeks at a time, emotions get bottled up and have nowhere to go. I get sucked into Facebook, and I keep scrolling down from one thing to the next, yelling at my computer as the posts get more and more insane. Mike Pence was just here in Minnesota, visiting patients who have coronavirus at the Mayo Clinic, and he went against their hospital policy and refused to wear a mask. It's like: "Really? Why? When did this country stop believing in science? How arrogant can you be?" Next it's someone posting pictures of people crowded together like sardines at a beach in California. "You idiots. Do you care about anyone but yourself?" Then it's the president saying he's doing such a great, marvelous, fantastic job, and oh by the way it might be a good idea to inject some kind of bleach or disinfectant as a way to kill the virus. "No thank you, but you go right ahead if you want to poison yourself." Then it's a militia group taking over a state capitol. It's doctors who have to wear garbage bags instead of gowns because we don't have any basic equipment. It's that we still don't have enough tests. It's how people keep saying that at least most of the deaths are people over seventy with preexisting conditions. "Oh, what a relief! Who cares about them?" It's some stockbroker or whatever saying the elderly are holding this country back from reopening, and maybe it's their patriotic duty to be sacrificed for the sake of the economy. "Sorry to be an inconvenience to your financial portfolio. Sorry I'm still breathing."

It enrages me. I was watching the news the other day, and one of the broadcasters started going through the obituaries in Minnesota and reading off the ages of people dying here from the virus. He was like: "Eighty-five, seventy, seventy-six, eighty-two. See? It's mostly just the elderly. These people would be dying anyway. It's really nothing we should all be this scared about." So what are you saying? Now I'm expendable? I'm holding us back? I spent my career working for the federal government at Veterans Affairs. I raised my kids by myself. I basically had to raise my ex-husbands. I marched and fought for women's rights. I volunteered for political campaigns. I pay taxes and fly a flag outside my house because I'm a patriot, no matter how far America continues to fall. But now in the eyes of some people, all I am to this country is a liability?

Sometimes, before I know it, I've been writing comments on Facebook posts for hours: "To hell with you then." "You idiot." "How dumb can you be?" "Moron." "Racist." "Selfish pig." "Idiot." "Idiot."

Everyone knows me as a kind person. I used to wear a peace necklace. I've gotten old enough that I just say whatever I think without any filter, but I don't always like what comes out. This isn't how I used to be.

There's a lot I don't recognize about what's happening now. This country is so completely different from the one I came into. My uncle was at the Battle of the Bulge the day I was born. I arrived right near the end of the war, and most of my life was American boom times. We were the leading country in everything when I was young. Nobody could touch us. My dad left for a while to work as a chef on the Alaskan Highway, and he traveled through Canada so we could carve a road two thousand miles over the Rockies in the dead of winter. We did whatever we wanted just to show that we could. We were invincible in this country. That's how it felt. I graduated from high school and started working when I turned eighteen, and within about a year I was earning more than my parents. That's how it went. It was up, up, up.

And what are we now? We're mean. We're selfish. We're

falling for all these conspiracy theories. We're stubborn and sometimes even incompetent. That's the face we're showing to the world. The great countries are supposed to rise up in a crisis, right? And what are we doing? It seems like some of these other countries almost feel sorry for us. New Zealand and South Korea beat this virus back in a few weeks, but we're basically getting numb to the fact that thousands of people are going to die every day. "Oh, it's just the old people. It's just those preexisting conditions." We've gone from ten thousand deaths to thirty thousand to a hundred thousand and some, so I guess we're still leading the world in that.

We can't get out of our own way. Are we shutting the country down or opening it back up? It's the states against the feds. It's conservatives against liberals. There's no leadership and no solidarity, so everybody's doing whatever they want and fighting only for themselves, which means everyone who's vulnerable is losing big-time. Minorities. Poor people. Sick people. Immigrants. Elderly. We're the ones who will die from this virus and the ones who will never recover. We've gone from taking care of people in need to thinking only about ourselves. That's the truth I'm learning about this country, even if I should have known it earlier.

I don't like feeling this way. Maybe somewhere in this we'll see a great lightning strike of American ingenuity. I doubt it, but maybe. There's no choice but to be hopeful. I'm staying alive and sitting in my house and waiting. Where else am I going to go? I'll be here.

Chapter 6

"Is this another death I'll have to pronounce?"

Michael Fowler, coroner of Dougherty
County, on the reopening of Georgia
Albany, Ga.—May 2020

I'm always driving, going back and forth between nursing homes, the hospital, and the morgue. All these roads around here should be empty if you ask me. But now I see people out running their errands, rushing back into their lives in the middle of this pandemic when we've got nothing solved, and it's like: Why? What reason could possibly be good enough? Sometimes I think about stopping and showing them one of the empty body bags I have to carry around in my trunk. "You might end up here. Is that worth it for a haircut or a hamburger?"

You start to think that way as a coroner, especially now. I get fed up. I know the governor told us we could go ahead and reopen here in Georgia. I understand there's political pressure and businesses are hurting and people need to work and get back to their lives. But sometimes I see these folks out and about and I start to wonder: Is this another death I'll have to pronounce?

My work never shut down in this pandemic. I've been busier than ever during all this. For six weeks now, I've been answering calls in the middle of the night, taking photographs at the scene, notifying families, trying to get those other family members tested. We've had more deaths in the last month here in Albany than we normally have in six months. This is a small community, but we've become ground zero for what they call community spread. Our greatest strength has been turned against us. We're close-knit, social people down here. We shake hands. We hug. We go to church and hold on to each other, and this virus takes advantage of that and keeps passing right along. We're killing each other now, literally. Our hospital is full, and they're opening up another emergency unit to create more space. We had two hundred more people test positive just last week. We've got so many sick people here in Albany that they're shipping some to Atlanta to find space. My cousin's up there now on a vent with this virus, and it isn't looking so good. Nothing's looking so good. We're right in the thick of fighting this thing, but all of a sudden the governor decides that now is the time to reopen? This is the moment when people really need to get a tattoo or make another trip to the gym?

I don't believe in getting hysterical. It doesn't do any good. This is a numbers-and-facts job. But we have numbers and facts right now that are screaming out by themselves.

I deliver an update to the town council each week on the death toll. I put on a shirt and tie and a mask, and I go downtown to give my presentation. My whole refrain in this has been: "Let's keep it below one hundred deaths in this count." At first, people thought that number was impossibly high. Crazy. I had people in this town telling me that I was adding to the hysteria. One hundred is a lot in a place where it seems like everybody knows just about everybody. But when I gave the update three weeks ago, we were at 49 deaths. The next week it was 67. Then 84. I gave my last presentation a few days ago, and we'd hit 102. By the time I got back to my office, the phone was ringing again with another call from the hospital, and within a few hours we were up at 107.

It's been so bad for so long that some people here say they're starting to lose count. Maybe so. But I never lose count. We have less than 90,000 people in this whole county, and close to 1,500 have already tested positive. All I do is keep counting.

I'd like to lie and say we were perfectly trained and prepared for something like this to hit—that all of us on the front lines in Albany were waiting and ready. I'm no rookie. I've counted and examined the dead and done forensics after all kinds of major disasters. I've had three back surgeries from the strain of moving heavy bodies. You get your floods and your tornadoes down here in South Georgia. But we're thirty miles from the nearest highway, and our airport has maybe two or three flights a day back and forth to Atlanta. You feel a little removed from what's happening in China or Rome or even New York. This isn't what you'd call a disaster epicenter.

The first alarm started going off in Albany on March 15. It was evening time. I got a call at home about a new case, and I went out to the residence. This was a forty-three-year-old Black female. At first, the process was routine. Usually, when the first responders are done, it's my job to pronounce the death and take a look at the body. I take pictures, look for any medications, talk to the family, fill out the paperwork, and decide if we need to do an autopsy. Sometimes the body gives away secrets. With an alcoholic, you might see a little clubbing of the fingers, and heart disease can show up as a swollen leg. But this time, I couldn't see anything. There was no evidence of foul play or anything suspicious. It was one of those mysterious circumstances. I went out and talked to the family, and they said this lady had been healthy for most of her life. I prayed with them a bit and then asked all my questions. I'm a pastor, and I like to put hands on people when we're talking, listen to their concerns, really get to know them. This family kept repeating her symptoms: high fever, aches, coughing, lots of trouble breathing. It sounded straight out of what I'd been hearing about on the news. I said: "Have you heard about the coronavirus?" None of them knew what I was talking about. I talked them through it. I said: "I think we better get her tested."

It was a hunch, really. Right away I had this bad feeling. There were twenty people coming and going in that house, grieving and paying their respects, and who knows how many of them were already infected. There was a rumor going around at that time that maybe Black people couldn't get this virus. We were dealing with all kinds of confusion and misinformation. Nobody in that home was wearing a mask or gloves or doing any social distancing. At that point, I wasn't either.

Nobody had told us to do any of that. I started worrying that maybe this one death was going to turn into a dozen.

I sent her body to be tested, went home and tried to get back to sleep, but I got two more calls later that night. One was a resident over at a nursing home, and the other was in the ER. Same symptoms. Both suspected cases. In one night, we went from zero deaths up to three, and that's how it's been going ever since. It explodes outward in waves like a bomb. One person has it and goes to church, and then pretty soon half the pew is testing positive. I console people on the death of their relative one week and end up pronouncing them the next.

I never thought this job could give me nightmares. I've been lucky that way, and it's probably why I've lasted so long. I worked at a rubber factory here in town until it closed in 1986, and as part of the layoff, they offered a few of us free tuition to school for mortuary science. It started off as kind of a joke. I'd never seen or touched a dead body. Nobody grows up dreaming about doing this job. They had us living inside a funeral home in Atlanta, and that first night we got called to a wreck on I-20. A girl had flipped her car. It was a decapitation. We searched the highway with a flashlight, and after we finally found all of the body parts, I went home and I slept okay. That's when I first started thinking this must be what God wanted of me.

You see a lot of haunting things doing this work. I believe it has to be a calling. Most of the time, you're seeing natural causes, but an average week might have homicides, suicides, drug overdoses, car crashes, child deaths, drowning. Bodies get cold over time and harder to move. Muscles start to stiffen. We had a flood come through and wash out more than five hundred graves a while back, and I spent months sorting through bones and skeletons. I've worked twenty-three international disasters, from a plane crash in Guam to the tsunami in Thailand. I spent nine weeks sorting remains at the World Trade Center. I try not to remember individual cases. I do my examination and write down the circumstances and the cause of death, and then I tell myself: "Okay. It's gone. Let it be." I pray on it and I move on. I trained myself to do that.

But what I'm learning lately is, it's a lot harder when the body you're zipping up is a face you know or a face of someone you love. I've lived in this town for my whole life. This disaster came to *my* community. At least thirty of these victims are people I knew by name or considered my friends. Six of our preachers in this county have died. I've broken bread with all of those people. I've lost probably at least seven or eight more I know from church. Two neighbors. Three school friends. The probate judge who had the office next to mine at the courthouse. These are my contemporaries. I'm sixty-two. We've had thirty-six people here die in their sixties, and at least a dozen more who were younger than that.

I try not to count down the days or make projections about when all of this is going to be over. For all I know, with us opening back up, it's about to get a whole lot worse. The sad truth is, I'm almost getting used to it. It's starting to become routine. I always stand six feet away from the families now when I ask them my questions. I always wear my space suit whenever I'm anywhere near a body. I always take off my clothes and have my wife spray me down with Lysol as soon as I get home. We've run out of space in the morgue. The chamber of commerce has gone ahead and given me a tractor trailer with shelves to store extra bodies, which I might need depending on how reopening goes and how many more cases we get.

The phone calls used to wake me up at all hours of the night. Now I'm usually up waiting.

Chapter 7

"I have to be getting better, right?"

Darlene Krawetz, on what life becomes
when COVID-19 won't go away
Syracuse, N.Y.—May 2020

I've hardly moved from this couch in weeks, but right now my heart rate monitor says I'm at 132. That's double my normal. That's like if I'm climbing a mountain. How come? Nobody knows. Nobody ever knows. And why has my fever been spiking again? Do I need to go back to the ER? I'm on week 6 of this crap, and I still don't know if I'm getting better or worse.

I *have* to be getting better, right? I'm active. I'm healthy.

I'm only fifty-two, and I don't have any of those big preexisting conditions we keep on hearing about. I've been a nurse for thirty years, so I know how to take care of myself, and I've been checking my oxygen levels and monitoring my fever since the very beginning. I'm the kind of person who gets better. That's what I keep telling myself.

I'm up to 140 now. See? It's relentless. How long can a heart last like this? The palpitations come a few times every hour and go on for a minute or more. It's just banging, banging, banging, banging.

It feels like electricity is burning through my spine, and nobody can tell me why. It's like I'm sucking air through a straw. When I stand up, my ears start ringing until dizziness forces me back down. Every symptom is a whole new mystery. This virus is unpredictable and so, so violent.

It hurts too much to talk. I'm sorry. I'll try again later. I have to lie down and breathe through it. That's what they tell me to do.

．　．　．

The next morning:

My heart rate is back down now to 105. That's nothing to celebrate. That's still considered abnormal, but it's typical now for me.

I didn't use to be like this. This time last year, I was traveling around the rain forest in Costa Rica with my daughter. I'm a vegetarian. I've got grown kids in the military and a teenager at home, and we like to hike and kayak. I'm a positive, hard-charging person. My kids tell me I never learned how to sit still, but I think it's because I can't really afford to. I work two different nursing jobs to pay for our little house. I'm big into animal rescue, and before I got sick I was taking care of ten foster kittens, some of them bottle feeders, just two or three weeks old. They're medically fragile, but I like the hard cases. I'm a caretaker. That's kind of my nature.

I guess I might have gotten this virus somehow at the VA hospital where I work. We didn't have any confirmed cases there yet, but some of my patients were coughing, and the administration was being stingy about giving us the protective masks. Or my husband works at a grocery store, so he could have brought the virus home. Or my son might have had an exposure and given it to me. Who knows? It's one more mystery. I didn't even notice I was sick until another nurse at the VA asked why I was coughing. I figured it was allergies. Take some Zyrtec and get on with it. Hardly anybody in Syracuse had COVID at that point. What were the odds?

Then, after I tested positive, I thought I'd get a mild case. I told my husband: "Relax. I'm fine." I don't have diabetes. I don't have hypertension, COPD or anything like that. I thought I could stay home, push the fluids and vitamins, take care of myself, and be back at work in a few weeks.

Right away I started running a temperature of 103, and the Tylenol couldn't control it. I was shaking and cursing all day in bed, and the symptoms spread from there. I was head-to-toe exhausted. I wanted the whole world to let me be alone. I had equipment at home from my nursing work, and I started checking my vitals and saw my blood pressure shooting up. I've never had that. I'd get up to shower and start gasping for air. My son was also COVID-positive, and he ran a high fever and recovered within a week, but I kept on getting worse. Maybe because I'm older? Or maybe because a long time back I used to be a smoker? You can't get a definitive answer on anything with this. I started coughing to the point of throwing up. I coughed until I was incontinent. My lips were chapped from dehydration. I had headaches. Migraines. Heartburn. Rashes. I lost sixteen pounds in the first few weeks. I would lie down at night after taking melatonin and Benadryl, soaked in sweat and terrified of what might be coming next. What if I fall asleep and stop breathing? More Benadryl. More melatonin. Maybe try a Xanax. I'd lie there for hours, but it was nonstop insomnia. I'd turn the TV to Lifetime for a distraction, but I couldn't make sense of what they were saying. My mind was getting foggy. Sometimes I couldn't remember if my last Tylenol was five minutes or five hours ago.

One day my son needed money to buy groceries. I said I'd give him $80, but I couldn't count it out. I couldn't do the math. I handed him $50, then $70. I asked him: "Is this really happening right now or is this a hallucination?" He took the cash and counted it himself. He looked scared. He begged me to go get help.

I went to urgent care. The X-rays showed pneumonia, so they told me to go to the ER. I didn't want to risk a secondary infection at the hospital, and I knew they didn't have any magic treatment for this virus, but it was getting to the point where I couldn't take care of myself. There wasn't any choice. I was scared of the bill for the ambulance, so I asked my son to drive me to the emergency room door and drop me off. I wrote down my end-of-life wishes. They took me in a wheelchair. I told my son: "I love you. I don't know what's going to happen. I hope this isn't goodbye."

Damn it. I'm having another heart palpitation now. Hang on. Are these panic attacks? I never had them before. It feels like my heart is trying to jump out of my chest.

Oh God. Just breathe. Stay calm. It's up over 150 now. Something is really wrong with me. I can't do this anymore. I need to go rest. I need to figure this out.

.   .   .

A few hours later:

Okay. I'm a little better. It's hour by hour. I'm not sure I can handle it again if I have to go back to the hospital. That first stay lasted ten days, or at least that's what they told me. I couldn't tell the days apart. I had a little glass isolation room with a curtain they kept closed. There was nothing to see out the window except a parking garage across the street. I couldn't have visitors, and most of the doctors and nurses were afraid to stay in the room. It was okay. I don't blame them. I was too sick to talk and too scared to feel lonely. I appreciate what they did. They were honest about what they didn't know, and they tried. They kept throwing stuff at the wall to see what might stick.

They gave me a malaria drug, but it did absolutely nothing other than send my heart into overdrive. They gave me an antibiotic for pneumonia, but I still couldn't breathe without fifteen liters of supplemental oxygen. They tried vitamin C, magnesium, shots of blood thinner, baby aspirin, Tums, multivitamins, Xanax, cough syrup with codeine. It was like fixing a car when you don't know what's broken. The doctor said to me: "I'm sorry we don't know more. We're still learning how to treat this virus. You're the guinea pig."

They gave me inhalers and breathing exercises to do every hour, but my oxygen level kept dropping. They wanted to put me on life support, but I was afraid I'd never come off. One night when it had gotten real bad, the doctor came in and said: "We have a team ready to revive you in case you start to code.

We're going to watch you closely." Watching was all anybody could do. Then, one morning, my fever started to go down. Nobody knew why that happened either. But it stayed down for thirty-six hours, and they said I could go home.

Now I've got my oxygen on a long extension cord. I can make it to the kitchen or the bathroom if I'm feeling good, but usually I stay here in the den. My husband never caught it, so we're staying apart. He works as a manager at Wegmans, and if he got sick, we might be out on the street. The $1,200 stimulus payment from the government went to rent and hospital co-pays, and now we're burning through our savings. The car is about to get repossessed. I think we might lose the house. I want to work. I want to get back to my life and start putting things together. I don't want to be stuck inactive. My son is getting ready to graduate high school, and I should be planning a party. This isn't me. I can't just lie here. My husband says: "Relax. Slow down. Make peace with the situation." I can't. Are you kidding? Is this just how it's going to be?

I try not to think about it. What good does it do? It's better to distract myself by watching the news and checking my vitals, but they're always bad. One day I did manage to go outside and sit in the backyard for about twenty minutes, and that was a highlight. I've found some groups of people on Facebook who are sick with this virus, and they post about their symptoms and how they're doing. It's a mystery that we're all solving together. Why do my eyes hurt? Why can't I taste food? Some of these people have been sick ten weeks, fifteen weeks, and you start to get to know them, and then suddenly, poof, they stop posting and they're gone. Did they get better or did they die? It's a nightmare, but at least it makes me feel less alone.

My family stands in the doorway to visit sometimes, and other people text or call. "Are you feeling better yet?" It's like they're becoming impatient. They want to feel safe going out into the world. They want to believe everything's okay, and it's like my sickness is depressing to them. We managed to buckle down in this country for a little while, but now it's starting to get nice outside, and people need to work. The deniers and the protesters are coming out. One of my relatives went on Facebook and wrote that this whole virus is overblown, or maybe even a hoax. People want to minimize.

"Are you better yet? Why aren't you better yet?"

I don't know. I don't know anything. My brain keeps racing with unanswered questions. Are my lungs scarred for life? Is my heart damaged? Can I get sick again? Will I be hiking the Adirondacks this summer with my family like we planned, or will I be lugging this oxygen tank from the den to the bathroom for the rest of my life? My daughter is a nurse, and she's my health-care proxy, and lately I'm relying on her more and more to talk to doctors and sort through all this. I don't remember things anymore. I struggle with finding words. It's like I know the information is inside me somewhere, but I can't access it. It's buried.

I hate this virus. It's been two months of uncertainty and I can't take any more. Why are my legs burning? Why is my skin so hot? I need answers. I need help.

.   .   .

The next morning:

I'm back at the hospital.

My fever won't come down. The doctors say I have blood clots on my lungs and a mass on one of my organs. They're trying to figure it out. There's no timeline and no real prognosis. All I know is they're admitting me.

I don't want to live like this. I don't want to die. Please. Can anybody help me?

I've been crying my eyes out. The morphine is putting me in a fog. When will this damn thing let me alone?

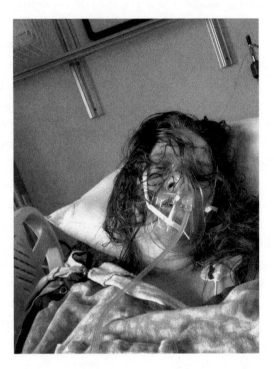

DANIELLE SWANN, Darlene's daughter

I was the one who was supposed to take care of her. That was always our deal, because she taught me how to be a caregiver. A few years ago when I was in nursing school, she wouldn't let me quit. I was living with her and I was struggling. The work was too hard for me, and I hated it. She would come home from the hospital and help me study for hours. She kept telling me: "I know you. Just get beyond the textbooks and you'll love this job. You were meant for this." When I finally graduated, I told myself I would pay her back. She would never have to go into a nursing home. I was going to be there to take care of her in the end, no matter what.

I tried. I tried the best I could, but maybe that's just an excuse. I don't know. There's nothing I'm at peace about.

We got sick around the same time. At first, it felt like something we were going through together. I live with my husband out in El Paso at Fort Bliss, but I had just gone up to New York to visit my mom, and about a week later we both came down with something. I couldn't breathe. I'm asthmatic, and that makes this virus about a hundred times worse. I turned blue. I called an ambulance many times, and it went on like that for a few months. My husband put me in a tub of cold water once and I woke up with him putting ice trays in there, trying to get my fever down. I would call my mom on the phone and compare all our symptoms. One day I would have it worse and the next day she would be worse. It was a race to the bottom. It was one thing after another. This virus thickens the blood and attacks the immune system, and different things started dropping. You don't have enough potassium, and calcium, and magnesium, and eventually your whole body starts failing. My immune system got so weak.

I was lucky. I started getting a little better, and then she kept having more bad days. It got to the point where I started putting off my own doctor appointments so I could focus on her. They give you a big book of medical facts in nursing school, and I'd go in there to look at her symptoms and figure out some of the medications. I'd call her doctors to get information, but sometimes they couldn't find the paper saying I was her medical proxy, or sometimes they would put me on hold for an hour and then they'd forget. The time change from Texas made it impossible. I'd ask my mom about stuff, but more and more she couldn't remember. My twin brother was at home trying to take care of her, and we would be up all night strategizing and talking it through, but he didn't have the same medical knowledge. I started thinking I should fly back.

I'm in the Army reserves, 74 Delta. We are trained to deal with pandemics and decontamination on a massive scale. In the Army, the joke about our specialty is that we're the car wash. We wear special suits and equipment, and we go in and wash everything and make sure it is safe. We were getting ready to possibly deploy to help with this virus, maybe in Asia, but I needed to be with my mom. I kept asking my brother: "How bad is she? I can always leave work if I need to. I have no problem even quitting if it comes to that." The priority was that I needed to be there with her. I just needed to know when. We talked about it, and I bought a ticket to come for a few weeks at the end of the month.

We thought there was still time. We kept thinking she was going to get better.

PAUL SWANN, Darlene's son

I didn't think she could die like this. Even once her fever hit 104 and her heart started racing out of control, my mind wouldn't go there. She was so healthy. She practically lived at the farmers market. The virus got into her lungs and her heart and her liver, and I still thought she was going to beat it. I was giving her CPR, and I was telling myself: This can't kill her. This won't kill her.

I caught myself doing the same thing at her funeral this week. I've never been to a funeral before, and it was all so weird. There was a prayer and a poem, and that was basically it, and I spaced out for a minute and started thinking she was back at home sleeping on the couch. I was like: I wonder if she took her morphine yet? I need to get back. I need to go check her temperature.

She was fifty-two. How am I supposed to accept that? I can't get it to sink in. My brain keeps on refusing.

It was mostly the two of us during the last few weeks. Her fever had finally come down and she was breathing on her own again, but she had blood clots in her lungs and a mass on her liver. It was growing faster than anything the doctors had ever seen, and they wanted to get her ready for chemo. They said her whole body was breaking down from the nonstop stress of fighting the virus since March. They said, "It's causing all these fires, and we have to put them out one by one," but it seemed like the fires were everywhere.

She wanted to come home until she started chemo. She had a fear of hospitals, even though she was a nurse. I thought it was a bad idea. She needed morphine and blood thinners and so many other medications, and there was nobody else to watch over her. Her husband had to keep working at the grocery store, because that was the only paycheck coming in. My twin sister's in Texas, and my youngest brother is only a senior in high school, so this shouldn't fall to him. I'd never taken care of anybody before. I'd just finished college and gotten out of the Army. She'd always taken care of me. I was scared of what might happen if I was in charge, but she was determined to be home.

She was on the couch in the den, and I would give her ice packs or help her change positions. I put her medications on a schedule. I cooked her au gratin potatoes and asparagus, but it took her maybe three hours to eat a little bit. She wouldn't drink the Pedialyte. She was having headaches and disorientation. She moved from the couch to the floor because she said it felt better to lie against something hard. I'd go try to nap, and if I got by myself, I'd start thinking about how it had been just a few months before. We'd go rock climbing together or shoot off fireworks by Lake Ontario. We'd road-trip to Canada because she wanted to try this certain kind of vegan food. She had this unstoppable energy. Her mother died in childbirth with her, so she had to fight and scrap from the very beginning. It was foster homes, abuse—she dealt with a lot. She knew how to push through. There was no way she wasn't going to beat it.

But then I'd hear her in the den mumbling or groaning or talking real low. It didn't sound like her. It sounded like something guttural, an animal. I was physically sore and tense from the stress, but I'd force myself to go check on her every hour. I was afraid of what I might find. How much of her is still in there?

Then one morning, she woke me up before six and said: "Hurry up. Let's go. We're running late for the doctor." She seemed coherent, but she was agitated. She said if we didn't hurry, she was going to get fined for being late. It didn't make sense. I'd been keeping track of her appointments, and there was nothing on the schedule. It was before sunrise. No doctor's office was even open yet. I thought maybe the medication was clouding her thinking, or she was trying to rush back to her old life. She was always a busy bee from the second she woke up. She would buzz around the house doing ten things at once, and if you got in her way, you'd get stung. I told her: "Your doctors want you to stay home and rest right now. There's no appointment. If you need something, they'll come see you." I helped her back onto the couch. She said she was sorry for getting confused, but it started to get worse.

She'd try to leave the house every morning at like three or four o'clock, but she could barely move. She was a wall walker. She would grab the car keys and inch her way out of the house in her underwear and then collapse into a chair on the porch. She was itching her legs really bad, and she started to get small infections because she was scratching past her layers of skin. She kept talking about how she was late for work, she needs to go work, she's going to get fined, she's going to get arrested. She didn't want to take her medication. She was refusing to eat. Sometimes she didn't know where she was. I called my sister in Texas, because she went to nursing school and she knows more about this stuff, and she told me to think of it like treating someone with Alzheimer's or dementia. She said: "You have to talk to her like a two-year-old. You have to comfort her and keep her company." I tried to mix her medication in applesauce and then take a bite to show her. "Look, Mom. We're eating together." I made a show of bringing the spoon up to her mouth. I told her it was going to be okay. I put the oxygen on her to calm her down. I tried to make distractions to get her to stop talking about going to work. I was firm and took away her keys so she'd stop trying to leave the house. I sat with her. I told her how much I loved her. I held her tight and cradled her like a baby.

It was too much. We needed help—a full-time aide. Her husband and I started calling the doctors and nurses a few times a day. "I'm not comfortable. I'm sorry, but I'm not equipped. What do I do? She's not listening. How do I take care of her?"

They sent a nurse out to evaluate. The appointment was for 9:30. I had to get her up and get her ready, but I was scared to wake her, because I never knew if she'd be a little combative or confused or trying to run out of the house. She said to me: "Paul, can you take me to the bathroom?" She was so helpless. I can't explain. It was gutting. Just looking at her was traumatizing. She was down to 111 pounds. She was losing weight in her face, her legs. She was getting that belly you normally see with starvation. She knew who I was, but it was like her eyes wouldn't focus. She had that thousand-yard stare. I tried to pick her up to take her to the bathroom, and she couldn't move her body. She could barely lift her arms. She was just dead weight. I said, "Can you roll over?" I didn't know if she could hear me. I said: "It's okay. You're okay. I'm going to go and get us some help."

I went to the porch to see if the nurse was there yet, and then I heard this weird gasping sound. I turned around and she wasn't blinking. It made my heart stop.

The nurse walked in right as I was calling 911. The emergency operator told us to move her to the floor, get her body straight. "You need to do CPR." The nurse couldn't help do the puffing, because she could catch the virus that way, so I ended up trying to give my mom the air. Her eyes were gone. It wasn't her. She wasn't in there. The ambulance came and they probably worked on her for twenty minutes. The sheriff was there. The house got crowded and we got pushed back. They started taking her out on a stretcher and I was looking to see if they had covered her face or anything. We kept asking if there was a pulse, is she breathing, but they wouldn't say.

They pronounced her right away at the hospital. The doctor told us it was a pulmonary embolism—a blood clot that got trapped in her lungs. He said it was something you see a lot of with COVID. He called it a sudden death, but it didn't really feel that way. She'd been getting carved up a thousand different ways since she got sick back in March. I went to the hospital room and I sat with her body for a long time. I kept her company. I tried to think of something to say, some last words.

We wanted an autopsy. My sister flew in from Texas, and we started calling around late that night, getting transferred from one medical person to the next, trying to make the arrangements. We needed answers. This whole thing was a mystery. I had gotten the virus and basically had a bad flu for two weeks. My sister got it and recovered. Nobody else in the house got sick. So what happened to her? Why did she die? Was it all because of the blood clot? Did the pneumonia factor in? The morphine? The mass on her liver? The weight loss? Some kind of vitamin deficiencies? All of this not knowing had been driving her crazy for two months, and it almost felt like we owed it to her. What happened? Why? How come she never got better?

The hospital and the medical examiners were putting us in circles. The autopsy cost $3,000, and nobody had that. We started scrambling together the money, but they were reluctant to do it because of the risks to their staff with exposure to COVID. It went on and on. Her husband is Muslim, and he wanted to respect the body, wash the body, bury it right away. He was being real accommodating, respecting our wishes and putting us first, so he pushed the funeral back three days. But time kept passing, and we weren't getting anywhere.

Eventually, the funeral day came, and we just had to accept it: It's never going to make sense. Nothing is solid. We'll never really know. She caught the virus and it kept assaulting her body for months until it was too much to handle.

So, we let it go. We let her be.

Chapter 8

"What happens if they run out?"

Johnny Rivero, on his first time standing in a food line
Brandon, Fla.—May 2020

I've never done anything like this before in my life. I'm not one for freebies. I don't even know how it works. I've been standing in this line now for a few hours, and it's barely starting to move. I'm not complaining. It's a blessing to be here. I'll wait all day if I have to, because this virus has left me with no other choice, but what happens if they run out of food?

They told us to be here this morning at ten, so I got here at eight. That's what they teach you to do in the military: "Be punctual. Take personal responsibility." These last few weeks,

I haven't been able to sleep all that much anyway, so I thought I'd get up and beat the crowds. But the parking lot was already full when I got here, and now people are lining up around the block. I saw a video on the news about traffic jams to get a handout at a stadium or a fairground somewhere, but this is just your basic little suburban church. It's probably one of a hundred food pantries that are open every week around Tampa, and I bet they've got three or four hundred people waiting.

They're promising us we'll all walk away with food. They're telling us to be patient, because I think this is double or triple their normal crowd. A lot of people in line are first-timers like me. Some of us have been praying together. A lady up front in a wheelchair has been crying about the heat. There's another guy who's kind of barking at his kids, but mostly the line is calm. Everyone is six feet apart. A few people brought chairs to sit on, and some are carrying empty shopping bags. Was I supposed to bring something? I didn't know how to prepare. Do they want you to show ID or give some kind of explanation of why you're in need?

I tried my best to avoid this. It's not that I'm ashamed to be here—not really. I've been on the giving end plenty in my life. I'm not stealing, and my family needs to eat. But I've lived fifty-seven years on this earth without asking for a handout, and it doesn't feel natural to start doing it now. I held off for as many weeks as I could. I prayed. I looked up directions to come here the last two weeks and then told myself we didn't really need it. I'm stubborn, and I keep giving myself these little pep talks. "You've lost jobs before. You can always find another." "We don't need a house phone." "We'll pay that bill later." "Turn off the AC and put on a fan." I've lived in some off-the-grid-type places, and I'm used to managing without savings, but there's no faking it when you don't have enough food to eat. You can't talk your way out of that problem. That's basic need. That's survival.

I don't mean to sound dramatic. Nobody's been starving in my house. I don't want to make it out to be worse than it is. We still have canned beans, rice, a few leftovers in the fridge. We have spaghetti to feed my three-year-old grand-baby, but even she's starting to get tired of spaghetti. I don't know. Maybe I could have waited it out another week or two, but what good does that do? This pandemic is never-ending, or at least that's how it's starting to feel.

We held steady until the end of March, and then it all fell apart. My whole family got laid off the same week. My wife worked at Party City, and they were ordered to close. They're trying to reopen again soon, but they're only bringing back a quarter of their staff, so it doesn't look like she'll be going back. Every business I know is shrinking. My daughter does market-ing, and nobody's selling or buying anything, so that job went away. I worked at a college, and now the campus is closed and their classes went online. I was doing building maintenance for $12 an hour. It wasn't much, but it was good for me. I could use my hands, do something useful, and drive over to spring training after work to see my Yankees. And then—wham. We went from a three-income household down to nothing.

The only place I've seen that's still hiring is Amazon. I can't do that. I did twenty years in the Coast Guard, and my body is falling apart. I've got bad arthritis, and I'm worried about getting sick. I have six relatives that live in the same apartment in New York, and they've all gotten this virus. It's ugly. I wear my mask everywhere. I try to stay inside. My wife and I have been hoping to get unemployment, but the website shuts down and there's no way to get through on the phone. It seems like the whole state's suddenly out of work. Our unemployment rate went from like five percent to fifteen percent, and the system is jammed. Sometimes you call the unemployment number and they tell you the average wait time is like six or seven hours. Then after you've been pressing the phone to your ear all day, you might finally get through to talk to a robot, who transfers you back to a real person, who tells you to print some paperwork off the website that isn't working, and that's usually about the time the call accidentally disconnects. We've been waiting to get food stamps. We've been waiting for another government stimulus. Yesterday, I waited almost two hours for my free medicines at the VA. I'm telling myself the system's overwhelmed and this is the time to be patient, but I can't make any more excuses to my bank account. It's hold for this, wait for that. We had a little savings but now that's all gone. I tried selling off some of my Yankee memorabilia, but anybody who might be interested is also laid off. What can I do to come up with a few dollars?

I can see a few people walking back out of the pantry area now. They're carrying bags and boxes. This must be the six a.m. crowd. I see cauliflower, canned soup, lots of potatoes. Those people look happy. The line is moving again. I've been here all morning, but I'm getting close. The organizers are telling the people in front of me to wait and they'll bring out their food.

I've been taking a few pictures in line to document this whole thing. It sounds weird, but this is historic. I document things. That's what I like to do. My wife told me: "Maybe now you finally have time to put together your book." I have thousands of pictures from my life. I was stationed all over the country. I did missions to the South Pole and hurricane cleanup in Puerto Rico. I volunteer at the VA with Spanish and Latino residents in the nursing home, and we like to sit and share stories. You start getting older, and your memories and experiences are what you have left. This is a story. That's how I'm choosing to look at it. This is a new experience that I'm living through, and it's humbled me.

I'm tired of listening to people blame and complain and put this all on the president. He's our commander in chief, and we're fighting this virus. We should be uniting and supporting, but we're tearing each other down. I don't understand it. We're lucky to live in this country. The bottom can fall out of your life, and there are people ready to help you out and give you something. That's the feeling I'm trying to focus on. It's gratitude.

Hold on. I'm at the front of the line now. They're coming toward me with a full cart, and it's got three bags of food and one box. "Wow. Thank you. God bless."

They gave me a lot. I've got plantains, blueberries, lettuce, fish fillets and fresh yellow peaches. There's enough bread for me to share with the neighbors. They even gave me a mango and a big pineapple. They just hand you the food and it's yours to take to the car—no guilt, no questions. They thanked me for coming. They *thanked* me. Can you believe that?

It's such a relief. It's more than I could have hoped for. We can probably make this last for three or four days, and they say I can always come back again next week.

LISA SCALES, president of the
Greater Pittsburgh Food Bank

Have you seen the pictures? I mean, my goodness. I remember looking at those images and thinking: This is the worst kind of history. Children for generations will be looking at this in their schoolbooks and thinking: This happened here? In America? Really? I remember our first food distribution after the pandemic hit and the economy shut down, and it was just car after car backed up on the freeway. The police chief told me the traffic jam went on for three or four miles. It was a line of endless needs. The desperation, the fear I saw from people that night—it humbled me. It scared me.

I'll be honest with you. We weren't prepared. How could we be? I never imagined I'd see a day in this country when so many people were pushed over the edge. We have the biggest food bank in Pittsburgh, but none of us had ever planned for a night like that. As soon as the pandemic hit, our phones started ringing off the hook. We went from ten calls a day to three hundred. We had people who started showing up at our warehouse, begging for a few boxes of food. I told our staff: "We have to set up a major distribution as fast as we can." We did it in our parking lot that first week after the governor shut down the state. We were planning on our biggest crowd ever—maybe five or six hundred people. It was pouring down rain. I went outside three hours before we were supposed to open, and the parking lot was already full.

For the first few hours, I was trying to emotionally connect with people one-on-one like we're supposed to do. I stood outside and loaded food into their cars. People were afraid to roll down their windows because of the virus. They were terrified and they were ashamed. It was a very humbling experience: seniors, people on oxygen, young moms with kids. A lot of people had just been laid off, and we had hundreds of first-timers. There was a woman with a gas can in her passenger seat, and she asked if we could give her a little more gas so she could make it back home. People honked and waved after they got their food. There was one lady who cracked her window and handed me a note. It said: "Thank you. Tonight we won't be so scared."

It went on for car after car, hour after hour. I started to go to a dark place. I've been doing this work for forty years. You start thinking: What problems have we solved? How can we have this many vulnerable people? Half of this country lives paycheck to paycheck. There's no safety net. We were set up to get demolished by an event like this. I got kind of nonverbal and started just loading food into cars. "Backseat or trunk?"

We started to run low on boxes of food. I went into the warehouse for more dry food and took whatever we had left. I called the police chief to see how many cars were still waiting. I said: "Does this line ever end?" He said we had hundreds of cars still backed up on the road. The traffic stretched all the way to the next town.

It was almost midnight when we finally ran out of food. How do you tell people who've been waiting in line for hours that you've got nothing for them? I kept tearing up. I apologized. I said: "Please. Come back." That part haunts me. I mean, we gave away five million pounds of food that month. We started doing bigger distributions at the Steelers' stadium, the amusement park, and the airport. But the thing I keep thinking about is the people we missed.

Chapter 9

"I know it was me"

Francene Bailey, on passing the coronavirus to her mother
Hartford, Conn.—June 2020

They keep telling me it's not my fault, and I'd give anything to believe that. The doctor called after my mom went to the hospital and said: "Don't blame yourself. You didn't do anything wrong." The pastor said basically the same thing at her funeral. "Let it go. You had nothing to do with this."

I know they're trying to make me feel better, and I appreciate that, but it's a lie. I had everything to do with it. This virus doesn't just appear in your body out of nowhere. It has to pass from one person to the next. It has to come from somebody, and this time I know it came from me.

I keep thinking: What if I'd stopped going to work when the first people started to get sick? What if I didn't live with my mom? What if I'd stayed upstairs in my room like I'd been doing all week? What if I'd kept my mask on? What if I'd turned away when she reached out to hug me? We only had close contact that one time, and it barely lasted a few minutes, but that was all it took. This virus is so contagious. A week later, she was in the hospital. Ten days after that, she was gone. That's the timeline I have to live with, and it points right back to me. I got sick and then she got sick. I lived and she died. How am I supposed to let go of that?

The thing is, I was trying so hard to be careful from the very beginning. It's not like I was one of those people who didn't pay attention. I work at a nursing home. I knew how fast this virus could spread. As soon as a few of the residents started spiking fevers in March, every single one of us out on the floor knew what was happening. The supervisors kept trying to say it was just an outbreak of the flu, but the flu doesn't sound like that. The whole floor was coughing. We went to the supervisor and said: "Please, we need some better protection." They didn't have the right supplies, and they were trying to downplay it. They said there was a shortage. They gave us each a paper bag with one mask, and it wasn't even the right kind. They said we had to wear it every day until it was torn. We started holding those masks together with string and tape.

I wanted to quit and stay home. I thought about it. But we've got three working adults in this house doing hourly jobs, and it takes every dollar from each of us to make it work. I've got maybe like $200 in savings. How long is that going to last me if I quit my job? I kept putting on that one mask and going back into work, even though it scared me.

You can't social-distance when you're a nursing aide. There's no way. I work on the lockdown unit, with patients who have dementia and Alzheimer's. These people need a lot of help. We feed them. We wash them. We do everything. It was only a matter of time before I got sick. I came home one day with a slight headache, and then I started to cough. My mom said to me: "This doesn't sound like your usual sinuses. I think this is different."

I told her not to come too close to me. She was healthy for a seventy-year-old lady, but I wanted to be safe. I called off work. I moved out of the bedroom I share with my daughter and her father so I had my own space upstairs. I started drifting away from everybody. I didn't know for sure if I had corona yet because I was waiting on the test, but I had a pretty good idea. My five-year-old would stand outside the room for hours calling after me. She likes to cuddle underneath you. That's the kind of person she is. She kept banging on the door. "Mommy. Mommy. Let me come in. I need to see you." I begged for her to go away. It hurt my heart, but I knew I had to do it. My biggest nightmare from the very beginning was that I might get someone else in my family sick. I kept telling my daughter: "Please, baby. Pretend like I'm not here." She couldn't do it. She wouldn't leave. She kept pounding on the door. Eventually, I had to stop answering and ignore her. I finally heard her walk away and ask her dad if I was dead. I sat in there by myself and I cried.

We're a tight family, and all of us have been living on top of each other in this house for twenty years. That's how we like it. It's the Jamaican way. I'm upstairs with my kids and my sister is downstairs with hers, and my mom went back and forth between the two floors. We share the bills and the child-care. We rely on each other. Some days, we might have fifteen people staying here, and my mom was always at the center. She retired from doing housecleaning at the DoubleTree in 2017, and ever since then she just wanted to rest. She wasn't a person of so many words, but she would sit in the kitchen all day, watching and listening. She knew I had gotten something bad. She had ten children, and she's a caretaker. She's got her own cure for every sickness on earth. I had to fight her from coming into the room. She told me to steam my head with orange and lemon. She started making this tea for me to drink. It had turmeric, garlic, lime, honey, and ginger. She would put it in a cup and leave it at the door five or six times a day. She stood outside in the hall and called me on the phone to make sure I was drinking it. "Put the phone where I can hear it go down." When I lost my voice and couldn't talk anymore, she would stand out there and text me: "Did you drink it? All of it?"

Anytime I heard people moving around in the hall, I would never go outside. If I needed to leave the room, I waited until it was quiet. They say the average person gives this virus to three or four people, but I thought: This is going to die inside me. I drove myself to go get the test so that I could make sure I was positive. I drove myself to the hospital a few days later when I knew it had become pneumonia. I drove myself to the pharmacy to get all the meds even though I was hyperventilating so bad I could barely hold on to the steering wheel. I took Clorox with me every time I went to the bathroom and tried to sanitize behind myself, but sometimes it got too hard to stand. I would text my mom and my sister: "I sprayed the bleach but I had to leave it. Let it sit until morning and I'll come finish the job."

At night, I was like a caged animal. I couldn't breathe, and

lying down made it worse. I was running a fever and the doctor said I needed to keep checking my oxygen. I had coughing fits that lasted for more than an hour. I drank so much
cough syrup that my body started to smell like it. I would pace
in the bedroom all night, from the wall to the doorway and
back, counting steps and watching the clock, trying to make
it to morning. I was scared and I was lonely. My mom has the
bedroom right underneath mine, so she could hear my feet on
the floor, and she would call out in the night. "Are you okay?
Francene, you're scaring me."

One day my back and my neck were on fire and I couldn't
keep walking. I tried to lie down, and it felt like the whole
house was falling in on me. I couldn't breathe. It felt like I was
dying, and I started to have a panic attack. I took off running
because I wanted to find air. I went downstairs, and I kind of
tripped over my slippers. My mom heard me, and she came to
the bottom of the stairs. I was gasping and sobbing. I couldn't
talk. She told me: "Stand still. Breathe. Take off your mask
and let the air in."

I pulled my mask down around my neck for a minute, and
she held me. She was trying to calm me down. I needed it, and
she needed to help. Our faces were touching. I was breathing
on her. I wasn't thinking about anything. I leaned on her until
I was calm again, and then I put my mask back on and went
upstairs. I tried to forget about it. It was only two or three
minutes. I tried to tell myself it would be fine.

A few days later, I heard her start to cough downstairs in her
room. It was nighttime, and I leaned against the floorboards
to listen. I said, "Oh God, no. No. Please, Jesus, don't let her
be sick."

But I already knew. She sounded exactly like me.

She had diabetes, so maybe that's why it went downhill so

fast. I don't know. Every symptom I got, she got it twice as bad. She was so out of it that she stopped taking some of her other medications. I talked to her once over the phone when she was at the hospital. She had a mask on her face to get oxygen, and the doctors didn't want her to do much talking. They were trying to get ready to put her on a ventilator. I told her she needed to listen to the doctors. I told her it was all my fault. I told her I was sorry. I didn't have my voice back, so I was kind of whispering, and I'm not sure if she could hear me or understand me. She said: "Don't worry about me. Focus on yourself. Are you drinking the tea? Please, drink the tea."

I was still in isolation in the upstairs bedroom when her doctor called again. The Department of Health told me to stay up in that room until three days after I stopped having symptoms. It was seven in the morning, and I was winded from taking a shower. Sometimes it took two hours for me to recover from the shower and get dressed. The doctor said they were doing chest compressions on my mom, but she wasn't going to make it. I started squeezing onto my legs, trying to find something solid to hold on to. The doctor said it wasn't my fault—that the virus could have come from anywhere. I told him: "What do you mean? She never even left the house. It was me. I know it was me, and I killed her." I threw the phone. I was so lost and so angry. I didn't want to hear it.

The phone kept ringing. People started coming over to grieve, and I heard them downstairs, crying and consoling each other. A few of them knocked on my door. They were worried. Nobody was blaming me. My nineteen-year-old stood in the doorway and talked to me for like an hour, telling me it was okay, trying to get me to come out. I told him: "No. I'm not getting near anybody." I closed the door and stayed upstairs by myself.

I called my bosses at the nursing home and just lashed out at them. I said: "You all messed me up, and I killed her. You weren't prepared for this virus. You didn't protect us. You let us get sick, and you just destroyed my whole world." I hate myself for going to work when I knew it was bad. Why couldn't they have taken care of us? Why couldn't they have paid us more? I hate that I needed that money so bad that I didn't listen to my instincts. I don't care anymore. I'll go broke before I go back to that place. It's too much to bear.

It's been almost a week since her funeral, and I'm still afraid to go outside. I'm scared to be within ten feet of anybody. I start shaking whenever I walk out the door. What if I catch it all over again, or what if I can still give it to someone else? I feel like I'm a walking bottle of poison, and I'm scared of what I might do to people. The doctor told me that feeling's not factual, since they cleared me as recovered. He says it's paranoia and anxiety. He wrote me a prescription and told me to take one or two tablets every time I leave the house, but I've decided it's easier to stay here. If I'm by myself, nothing else can go wrong.

Chapter 10

"We're all starved for hope"

Ian Haydon, on the trial and error of being
injected with a COVID-19 vaccine
Seattle—June 2020

I track the numbers along with everyone else, and it keeps getting more depressing. Forty thousand new cases each day. Fifty
thousand. Seventy thousand. How high can we go? There are
scientifically proven ways to fight this virus, and we've failed
at every one. We don't have enough tests. The testing we do
have is way too slow. We have no clear leadership. Millions of

people can't afford to stay home, and now we're throwing fits about social distancing or wearing masks.

It's like the United States is down to one solution that might save us from ourselves, one final bailout—the vaccine. Where's the vaccine? When can we get it? That's what the entire country is waiting on, so it feels strange that I didn't have to wait.

I was one of the first people in the clinical trial for Moderna. There are forty-five of us in Phase 1. This type of vaccine is a new technology, and it had never been given to humans before, so it's a lot of unknowns. Sometimes I hear people discussing a vaccine like it's some guaranteed silver bullet that should be ready to arrive on demand. We're all starved for hope. I get it. But this isn't magic. It's science, which means protocols and phases and data to collect. There has to be room for trial and error. That's part of what they're learning from me.

For the first month, my experience was very standard. I work at a biotech institute in Seattle, so I follow science at a daily level, and I learned about the vaccine trial on our office Slack. A co-worker posted a link with the sign-up form. He said: "This is crazy—they've gone from sequencing a vaccine to starting a human trial in about a month. If you've heard about the vaccine that might be ready within a year, this is the leading candidate." They were looking for healthy people between eighteen and fifty-five to be the first ones to get it. I'm twenty-nine, and lately I've been marathon training. My health history is so incredibly boring. I take a multivitamin once in a while, or maybe a Tylenol after I go running. That's basically it. I've had nothing happen to me, ever, so I thought I might be a good candidate. I filled out the form without really thinking about it, and a week or so later they called me back. I went in for an initial checkup to see if I qualified, and they took eight or ten vials of my blood and sent it off to NIH.

Eventually, they offered me a spot in the trial and gave me a consent form. It explained that I'd get two identical doses of the vaccine spaced twenty-eight days apart, and then I'd have a bunch of follow-up appointments to check my immunity and see how it worked. It said: "You might have adverse reactions. There may be unanticipated risks."

I talked it over with my girlfriend and my family. The pandemic was already crushing Seattle at that point. I have a great-grandfather who died of the Spanish flu in 1918, when he was only twenty-three years old. I put my trust in scientists to come up with solutions. I believe in experiments. I had a biology teacher tell me once that if you know the results before you start, you're not doing real science. I made my peace with the unknown. Here we are at this unprecedented moment for science, where you've got more than a hundred vaccine candidates racing to solve this virus. They're talking about getting a safe vaccine ready within a year, which is about five times faster than it's ever been done before. In this awful time, it was a chance to be a part of something truly historic. It wasn't a hard decision. I said yes and then signed all the forms.

They broke us up into three groups, and I was put in the study group that was getting the highest dosage of the vaccine. It was ten times more than some people got. I remember sitting in the car before my first vaccination shot for an extra few minutes to gather myself. I don't like needles. The exam room was really sparse, and I tried to distract myself by talking to the nurse and the doctor. I closed my eyes and turned the other way when the vaccine went in. In my head, it was this big, landmark occasion, but the reality was five seconds. It was nothing. It was any typical flu shot. My upper arm was a tiny bit sore. They gave me a thermometer to check my temperature and a diary to write down my symptoms, but I had no side effects and nothing to say until I came back for my next injection after twenty-eight days.

The first time was so easy that I wasn't really nervous for the next dose. The clinic experience went exactly the same. The first difference I noticed after the shot was arm pain. It came on hard within about an hour of the second shot. Then, at like ten-thirty that night, I was getting ready for bed and I started having chills. I was wearing sweats and still shivering, and I woke up all night with a whole bunch more symptoms. Nausea. Headache. Muscle pain. I took my temperature after midnight and it was 103.2. I was sort of loopy. They had given us a twenty-four-hour call line as part of the trial to reach the nurse, but I think some part of me was like: What is this going to mean for the vaccine? I didn't want to impede the progress. My delirium goal was to sleep it off.

At four in the morning, my girlfriend finally called the hotline, and one of the doctors in charge of the study met us at urgent care. The nurses were wearing those space suits for protection, and they gave me an IV for fluids and Tylenol for fever. They tested me for COVID and a bunch of other things with a full viral panel, kind of ruling out the possibilities. My sense, in piecing it together, is that I had an immune overreaction to the vaccination because of the high dosage. Basically, my immune system went into overdrive. The doctor told me: "It's good you called. This is exactly what we wanted you to do. We are going to learn from this information, and millions of people are going to benefit. This is the whole point."

My fever was already coming down, so I went home and slept until about noon. When I woke up, I realized I was really nauseous. I went to the bathroom to throw up, and when my girlfriend came to check on me, I was fainting. She caught my head so I didn't hit the floor. I woke up laid out on the ground, and I was trying to piece together what had happened and where I was. She called the hotline again, but by then I was lucid. I drank some fluids and rested on the couch.

It's probably the sickest I've ever been, but I was back to normal the next morning. I was fine. At no point did I think I was dying or anything. Even taking too much Tylenol can make you sick, you know? I don't want it to sound like more than it was. I'd hate for—I don't know. It's complicated to talk about. I worry it will get twisted by the little army of anti-vaxxers. We're living through a low point in confidence in scientific institutions, and all the disinformation scares me. People have asked: "Don't you regret signing up for the study? Didn't this give you doubts?"

It's like, "No!" For me, it's more evidence the process is working. This is what's supposed to happen. Two other people had negative reactions from the high dosage that lasted less than a day, and then Moderna announced they were discontinuing the high dosage as they moved into Phase 2. Nobody else will ever get as much of this vaccine as they gave to me. I guess that's my small legacy on whatever this vaccine might become. The lower dosages of the Moderna vaccines are producing antibodies, so why inject more? It's a trial. You learn and adapt as you go. They're still monitoring all of our blood work for Phase 1, and I feel perfectly fine. They're like vampires at this point. I've been back to the clinic probably four or five times since my second shot, and they're only interested in my blood. They fill a few vials and send it off to the lab to see if my antibodies are sticking around, or if anything screwy is happening with my immune system. Phase 2 is 600 people, and all of them have already gotten the vaccine. Phase 3 starts this month with 30,000 people, and that's across all the age ranges. A dozen other vaccine companies are right in the middle of doing the same thing.

This whole process is like walking over thin ice and looking for cracks. Step out a little further, test the ice, shift directions, keep inching ahead. Maybe you move a bit faster, but you don't close your eyes and sprint. You need to be sure. You have to know the ground is solid.

I don't know when that will be. That's the kind of stuff they won't tell us. I've heard people say Moderna is aiming to have 100 million doses this fall. I hope that's true. I hope science triumphs and the vaccine arrives later this year to bail us out. But all I know for sure is what it says in the paperwork they gave me. Phase 1 of the Moderna trial is scheduled to last fourteen months, and I have follow-up appointments into next June.

Chapter 11

"Heroes, right?"

Anthony Almojera, a New York City paramedic,
on the injustices of COVID-19
New York City—June 2020

Nobody wants to know about what I do. People might pay us
lip service and say that we're heroes, but our stories aren't the
kind anyone actually wants to hear about. Kids in this country grow up with toy fire trucks, or maybe playing cops and
robbers, but who dreams of becoming a paramedic? That's
ambulances. That's death and vulnerability—the scary stuff.

We're taught in this culture to shun illness like it's something shameful. We'd rather pretend everything's fine. We look the other way.

That's what I see happening now in New York. We just had 20,000-some people die in this city during the spring, and now the weather's getting nice out, and already crowds are lining back up outside restaurants and jamming into bars. This virus is still out there. We respond to 911 calls for COVID every day. I've been on the scene at more than two hundred of these deaths—trying to revive people, consoling their families—but you can't even be bothered to stay six feet apart and wear a mask, because why? You're a tough guy? It makes you look weak? You'd rather ignore the whole thing and pretend that you're invincible?

Some of us can't stop thinking about it. I woke up this morning to about sixty new text messages from paramedics who are barely holding it together. Some of them are still sick with the virus. At one point we had twenty-five percent of EMTs in the city out sick. Others are living in their cars so they don't risk bringing it home to their families. They're depressed. They're emotionally exhausted. They're drinking too much. They're lashing out at their kids. They're having night terrors and panic attacks and all kinds of outbursts. I've got five paramedics in the ground from this virus already and a few more on ventilators. A lieutenant over in Queens just shot himself with his father's gun. He'd seen enough. Another rookie EMT committed suicide last week. He was having trouble coping. He was a kid—twenty-three years old. He won't be the last. I have medics who come to me every day and say "Is this PTSD that I'm feeling?" But technically PTSD comes after the event, and we're not even there yet. It's ongoing stress and trauma, and we might have months and months to go.

Do you know how much EMTs make in New York City?
We start at $35,000. We top out at $48,000 after five years.
That's nothing. That's a middle finger. It's about forty percent
less than fire, police and corrections—and those guys deserve
what they get. But we have three times the call volume of
fire. There are EMTs on my team who've been pulling double
shifts in a pandemic and performing life support for sixteen
hours, and then they go home and they have to drive Uber to
pay their rent. I'm more than fifteen years on the job, and I still
work two side gigs to afford a basic life in this city. One of my
guys does part-time at a grocery store.

Heroes, right? The anger is blinding.

One thing this pandemic has made clear to me is that our
country has become a joke in terms of how it disregards working people and poor people. The rampant inequality. The rac-
ism. Mistakes were made at the very top in terms of how we
prepared for this virus, and we paid down here at the bottom.

It started for us around the middle of March when the call
volume began to spike in the poorer neighborhoods. The stay-
at-home order in New York hadn't even gone into effect at
that point. Trump was telling us he had everything under control. The mayor was saying we had great health care here in
New York, and we wouldn't get hit as bad as other places, so
we should keep on going out to the movies. But for us on the
911 calls, it was wheezing, trouble breathing, heart palpitations,
cardiac arrest, cardiac arrest. This virus stresses out the heart
in a bunch of different ways. I'd look at our dispatch screen
sometimes and see thirty possible cardiac events happening at
any one time across the city, mostly in the immigrant neighborhoods. It felt like watching a bomb go off in slow motion.
You had time to see who was going to get hit and who had the
ability to escape. I saw in Manhattan, on the East Side, people

clearing out of the city to set up shop in the Hamptons or rent property upstate. The business class packed up their computers and went to work elsewhere. The white-collar group starting hunkering down and doing their work from home. Meanwhile, the rest of us were out here trying to keep this city functioning: your janitors, your bus drivers, your bodega workers, your sick and your poor. We didn't have any choice to stay home. We were out here with the masses even when the virus kept spreading. We were like fish in a barrel.

I love this city, and that's why it hurts to see us get so exposed. I lived in Montana for a while on a ranch and learned how to do horse stuff. I've traveled to more than ninety countries, and that's how I like to spend my vacation time, but you can feel like you're traveling around the world just by being inside New York. There are ten thousand versions of this town, and in this job I get to see them all. This work is intimate. We grieve with people. We go into the crammed apartments when the stress level is sky-high. I've delivered sixteen babies since I started this job. You begin to feel like part of the city—like it's yours somehow and you're protecting it. The station I work at now is nine blocks away from the public pool where I went swimming as a kid. I've worked in Harlem and Queens and Midtown, and I love it all—different food, cultures, sounds. No two blocks are the same in New York. But during this pandemic, it got so quiet sometimes that all you could hear were our sirens. Emergency. Emergency. That's all it was. We're talking about the greatest, loudest, brashest city in the world, and nobody was out. Parts of New York felt like a ghost town.

The most 911 calls we'd ever had before all this was back on September 11, when the towers fell, and we broke that call record every single day during this pandemic for two weeks straight. My station is right in Brooklyn's Chinatown, so it's a lot of new Chinese immigrants, sometimes ten or twelve people living in a small place. They tend not to call 911 unless it's absolutely necessary, but they were calling. One woman was apologizing for bothering us while we were trying to get a pulse back on her uncle. The Dominicans and Puerto Ricans in Sunset Park got hit hard. Sometimes those families will pray over you while you're doing CPR. The Middle Eastern neighborhoods in Bay Ridge got hit. The African American communities, where hypertension is a big thing. The nursing homes in Far Rockaway. The housing projects in East Flatbush. A lot of times, we show up and the family is asking about how they can possibly pay for an ambulance. That's our demographic. We weren't carrying too many stretchers in and out of the fancy brownstones.

I'm a lieutenant and vice president of the union, so I cover a big area, and I mostly go to the big traumas. I grew up in Brooklyn, and I know every street in this city. It doesn't matter where the call is. I can whip it. I'm never more than about two minutes out. I had one guy with COVID who was talking to me in his fifth-floor apartment. He was breathing heavy, so we loaded him on the stretcher, and by the time the elevator hit the lobby, he didn't have a pulse. I went to another high-rise for an unresponsive elderly woman, and then I get into the apartment and I realize that two days before we were in the same place because her husband had dropped. Both of them died. We sometimes had four hundred emergency calls sitting on hold. The 911 wait times went up by like five or ten minutes all across the city. People were waiting hours for an ambulance on the more minor stuff. There are 4,200 EMTs in the city, and our turnover is terrible because of the low pay. Seventy percent of people leave us within five years. I've got twenty-six years on the job, which is longer than just about anyone else at my station has been alive. When someone has to inform the family that the CPR is over and we did every- thing we could, but their loved one is dead, you want that to come from somebody with experience. That sounds different coming from me than it does from a nineteen-year-old kid. I went from one fatality to the next. I pronounced more deaths in the first two weeks of April than I have in my whole career.

I got one call at the height of the madness, another cardiac arrest, and it was a Latin guy, young guy, unresponsive and passed out in a room with bunk beds. There wasn't enough space to work, so we dragged him out into the living room to start giving him CPR. This guy had no pulse. That's clini- cal death, but biological death doesn't come until about six minutes later. That's our window to bring you back. Those six minutes—that's why we do this job. When you go from no heartbeat to a heartbeat, there's no feeling like that in the whole world. And this guy was only thirty-one. He was strong, healthy. His mother told us he'd just gone out. As a medic, you hear that and your eyes start to get big. It's like, okay, maybe this is one we can save.

It was four guys and me. That's always the crew. The two EMTs were bagging him up to get oxygen in his lungs. The medics were starting to intubate and calculating the meds. Everything they can do for you in a hospital, EMS brings to you. We carry sixty medications. We hook up the heart moni- tor. It all happens so fast, and there's barely time to talk. Get your needle, put in the IV, pace it, shock it, check on the heart rhythms. The whole thing is like a symphony, and you have to know your part.

The team kept working, and I went over to get informa- tion from the mother. There was a little girl standing behind

her, seven years old, and it turns out that she's the daughter. I
usually let the family watch us work if they want to. I think
it can help. It speeds up the stages of grief, maybe gives them
a little more closure. This family told me he'd been sick four
or five days, but he worked at a bodega and he couldn't afford
to take off. He'd come home from work and collapsed a few
minutes later. Now I'm hearing this and I'm starting to get
upset. Here we're supposed to be this great society, and this
guy can't even miss one paycheck. There's no safety net in
this country. We've got a president lying to us about how the
virus is under control. We've got a mayor telling us to go to
the movies. We've got idiots walking into stores who are too
selfish to wear masks. Back during World War II, you had
everyone donating their nylons to the war effort, and now we
can't even get on the same page about if this virus is serious or
not even when thousands of people are dying every day. The
system we have is broken, and as a result this seven-year-old is
seeing her dad get CPR, and it makes me so mad. But we kept
working. I'm taking my turns doing CPR, and then this girl
is telling me that she shared the bunk beds with her dad. I'm
thinking, Please, please. Just let us get this guy back. Give us
one save. His skin color was starting to go gray. It didn't look
good. Usually, after twenty minutes, that's when I have to call
it. If there's no pulse, no electrical activity at that point, we get
the time of death and I pronounce it. We went to four minutes, six minutes, and then all of a sudden we got back a pulse.
The seven-year-old started crying. I told the family: "He's not
out of the woods yet, but we might have a shot here." We
rushed him into the truck and over to the hospital, and then
he died a while later.

I did fourteen cardiac arrests that day. I didn't save anybody.

The thing about being a paramedic is, you need to have some reservoir of hope. This job is the ultimate backstage pass. It can make you believe in the far limits of human possibility, but it can also suck the air right out of you. You see death, suffering, grief in its rawest forms. I've been shot at on this job. I've been beaten and cursed at. But then every year, we go to the Second Chance Brunch, and we get to meet some of the people we saved. There's no drug on the planet like that. There's no job that matters more. It keeps you going through the heartache and the terrible pay and all the rest of it. But then we came into this virus, and we weren't bringing people back anymore. The virus kept winning. It always ended the same way.

At least before all of this, when I lost somebody, I could sit there with the family, hold them, console them, listen to their stories. That's a big part of what we do—communicate and care for the family. You have incredible moments of compassion. Death brings out this common humanity. But the pandemic made it so they can't touch me. They can't see my face. I'm trying to tell them their mom just died, and they can't hear me, and they can't read the sadness on my face. I'm just some guy standing six feet away in a mask who failed to save the person that they love. Then I'm rushing over to the next call, because there was always another call, and that body might be sitting there in the apartment for hours until the police or the morgue finally has time to pick it up.

I'd go park the truck at the beach after a double and try to calm myself down and gather my thoughts. My friends say I need to take care of myself. It's like: How? I've gained weight during this pandemic. I don't sleep well anymore. Emotionally, I've been feeling a little numb. They teach you as a Buddhist that life is suffering, and I believe that. You have to stay in the suffering. You can't deny reality and turn the other way.

I've been in therapy seventeen years, and lately what keeps coming up in my sessions is that reservoir of hope. It's starting to feel more and more empty. The call volume is down right now, but I'm afraid it won't stay there. I don't have that much faith in what we are anymore. America is supposed to be the best, right? So why aren't we united at all? Why aren't we taking care of each other? The virus is hanging around, waiting for us to make more mistakes, and I'm afraid that we will.

Chapter 12

" 'No mask, no entry.' Is that clear enough?"

Lori Wagoner, on trying to enforce a
state requirement to wear masks
Oriental, N.C.—July 2020

We tried our best to be polite about it. I started out by framing it to customers like they were doing us this big favor: "Would you please consider wearing a mask?" "May we offer you a free mask?" "We sure do appreciate your cooperation."

I'll never understand what's so hard about putting on a mask for a few minutes. It's common sense. It's a basic way to stop this virus from spreading. It's even a requirement now in North Carolina. But this is a conservative area, and there are only nine hundred people in this town. We try hard to get along. I work in a small general store where we know a lot of our customers by name, and we didn't want to end up in one of those viral videos with people spitting or screaming at each other about their civil rights. We put a sign outside—an appeal to kindness: "If you wear a mask, it shows how much you care about us."

We found out how much they cared. It became clear real quick.

I'm sixty-three. I'm a lifetime asthmatic. I'd watch customers pull into the parking lot without their faces covered, and my whole body would start to tense up. Our store is on the Intracoastal Waterway, and people from all over the world dock in the harbor and come in here to pick up supplies. It's a big petri dish. I put a plastic shield up over my register, and a few hours into my shift it was covered with spittle. We'd have twenty or thirty people walk by the sign and come in without a mask anyway. I'd try to get their attention and point to the sign. It was a lot of "Hey, sorry, but I'm not going to wear a mask just because you ask me. You're infringing on my constitutional rights. This is a free country. I'm here to shop, so who's going to stop me?"

Then the local sheriff went on Facebook and said he wasn't going to enforce the state requirement because he didn't want to be the "mask police." So now what? I have customers who are breaking the law and putting my life at risk, and what am I supposed to do? I'm a freaking retail clerk. I ring up beer and boat supplies all day for ten bucks an hour. I don't want to deal with this. This is the last thing I want to be doing. The stress is slowly killing me. If I didn't need the money from this job,

I'd be home working in my garden or visiting my grandkids. I don't come into the store every morning looking to make some big moral stand, but when I see something that's wrong, I can't let it slide. I cannot shut up. I get stuck on things. That's my biggest downfall or my biggest asset. So, fine. I'll be your mask police. What choice do I have? I talked to my co-worker, and we decided we didn't feel safe having people walk around the store without their face covered. We agreed to hang another sign on the wall: "Thanks for wearing a mask. It's the most patriotic thing you can do."

That didn't stop them, so we kept adding more. "Please be kind to us." "We're here for you seven days a week, and we didn't create this situation." "Masks are required for anyone entering the store."

Maybe some people took it as a challenge. I don't know. But it kept on escalating. Most of our customers are supportive and respectful about it—maybe ninety, ninety-five percent. But on weekends, we get dozens of people from Charlotte or Raleigh who come to visit their boats. Those places are virus hot spots, and they come here to have a good time and maybe they're drinking. Some of them would see our signs, open the front door, and just yell: "Fuck masks. Fuck you." Or they would walk in, refuse to wear a mask, and then dump their merchandise all over the counter when I asked them to leave. I had a guy come in with no mask and a pistol on his hip and stare me down. I had a guy who took his T-shirt off and put it over his mouth so I could see his whole stomach. "There. A mask. Are you happy?" I had a lady who tried to tape a pamphlet on the front window about the ADA mask exemption, which is a totally fake thing. It's a conspiracy theory, but it's become popular here. She kept saying we were discriminating against people with disabilities. She kept telling me that masks were killing people. What? Why? How? None of what they say sounds logical. I can't even make sense of half the names they call me. They say I'm uneducated—uh, that's kind of ironic. They say I'm a sheep. I've been brainwashed. I'm pushing government propaganda. I'm a tool of socialism. I'm a Nazi. I'm suffocating them. I'm a part of the deep state. I'm an agent for the World Health Organization. "How do you like your muzzle?" "Is this going to become the new sharia law?" "Are you prepping us all to wear burqas?" "What's next? Mind control?"

The customer's always right. We grit our teeth and try to accommodate the customer. We offer them free masks, even though each one we give away costs us about a dollar. If the mask still makes them uncomfortable for whatever reason, we say they can wait outside and we will be happy to provide curbside service at no extra charge. If that somehow offends them, we apologize and suggest they shop somewhere else. Then it's "My rights, my rights. You're taking away my rights. Why are you trampling on the Constitution?"

My fists are clenched all the time now. I get nauseous before I have to go into work, because I'm always on edge. I wish this virus were glitter so we could actually see it, because in my mind, it's everywhere. I wear gloves to touch the merchandise. I wear gloves to hold the cash. I wipe down everything. I put a table in front of my register so nobody can come closer than six feet. I sanitize my hands so much they must be drunk. We had three new positive cases on the same day in this tiny little town, but people can't be bothered to put a piece of cloth over their face, even though that's a scientifically proven way to hold this virus back. Somehow they've decided a mask is a bigger threat to them than the virus. The sheriff's department is closed to the public because it has a bunch of positive cases with people on staff, but they still won't enforce the mask law. I don't get it.

It's bizarro world. One day I said to my co-worker, "I need to leave the store right now or I'm going to lose it. I'm going to explode." I went home and ended up taking twelve days off. It took me a full week to finally calm down. I had a dream that I was going around the store and physically moving people six feet apart, scolding them for not wearing a mask. I came back to work and decided I wasn't going to take it anymore. I handed out these laminated cards that say: "Mask Exemption Override—There is no ADA exemption for mask wearing." If a person refuses to wear a mask, I'm like: "Okay. Goodbye. Have a nice life, and thoughts and prayers if you get COVID." I'm sick of it. They're selfish. They're lemmings. I don't know if the virus will kill me or if it's going to be my rage. I'm all out of empathy right now. Sometimes I want to cut America into different pieces, and all these anti-maskers can live together in a society that refuses to believe in doctors or scientists or facts, and we'll see how it works.

A few weeks back, we put an orange traffic cone on the sidewalk out front to draw people's attention before they even come into the store. We taped up another sign. "No mask, no entry." Is that clear enough? That seems pretty clear, right? But this big, burly guy walked right past the cone and past all the signs, and he pushed the door open. I said, "Sir, can I help you?" I pointed to the signs. I pointed to my mask. He just shrugged. He was probably in his late thirties, and I'd never seen him before. He rolled his eyes and ignored me, so I knew where it was going. I came out from behind the register to try to block his path into the store. I said: "Do you have a mask you can put on? Otherwise, you need to leave." He shook his head like he couldn't be bothered, and he said he just wanted to buy a drink. I said, "Okay, that means I will get your drink while you wait outside, and I will bring it to the door." But he doesn't even hear me. He doesn't care. He's still moving into the store, and I'm trying to stay in front of his path and keep him from going down the aisle. He said, "Come on, lady. Move it or lose it. I just want water. I have an ADA exemption." I said: "I'm tired of this. That's not a real thing. I'm asking you to leave the store now."

He kept moving toward me, yelling, "ADA exemption, ADA exemption," and now my whole body was starting to shake. It was fear and so much anger. Why is this my problem to deal with? This maskhole? This COVIDiot whose stupidity is putting me at risk? This isn't what I signed up for. I'm a sixty-three-year-old lady and I'm trying to be the enforcer. I'm trying to corral this massive guy to the door, but he's not backing down, and he's getting more aggressive. He's screaming about his rights. He's calling me a Nazi. He's yelling at me to call the police. We're six inches apart. He yells out: "Social distancing! You hypocrite! Move out of my way." He's screaming all kinds of profanity, and I'm not being an angel. I'm screaming it right back. My co-worker was yelling for him to get out, and another customer started yelling, and finally he stomped around for a while and knocked into some things and then turned back outside.

We locked the front door and my co-worker and I went back into the storage room. We hid there for probably thirty minutes. We sat and we sobbed.

The next morning, I went to the hardware store to buy supplies. I can't handle the constant tension of this anymore. It's rinse-and-repeat with all these daily blowups, and I'm starting to get paranoid. All of this tension is getting worse, and sooner or later one of these customers is going to really fly off the rails. We installed a doorbell at the store so we can keep the front door locked even during business hours, and I've got pepper spray up at my register. This is my job now. At least next time I'll be ready.

NICK PEREZ, in Phoenix, Ariz.

There are certain people that shy away from confrontation, but I'm not one of them. I haven't worn a mask into a store for the last eight months. I don't care what somebody says I *have* to do. I walk right by all their signs. It's not even a law, okay? It's a mandate. That's not stopping me. I wouldn't say I go into places looking for a fight, but I'm sure as hell not going to walk away from one that's right on my doorstep. I'm tired of this. Who are you to tell me what I can and can't do? You think you can just take over my body and take away my rights?

Yeah. Okay. Go ahead and try.

It's not just about wearing a mask, even though I hate breathing in those things. It makes me claustrophobic. It's about protecting our freedom and rising above all this political propaganda. What happened to fighting for individual rights in this country? Just the basic right to operate a business has been stripped away from people in this pandemic. It seems like this country right now is willing to bend to whatever we are told. We are sheep. We follow along blindly, and this is just the start of what's to come in terms of government control. Can we force you to wear a mask? Can we force a vaccine on you? The government is seeing how far it can go in terms of taking individual rights. They're using this to float out a trial balloon. What other rights are going to be taken away?

I didn't start out thinking like this. I guess you could say it's been a gradual awakening. I was very intent on wearing a mask early on, before anybody else was paying attention to this virus. I'm somebody who goes beyond the mainstream media narratives and looks for my own news from some of the alternative sources. That's where the message isn't as controlled. You can find some good stuff. Back in early February, I saw videos online coming out of China with people getting sick. I bought $2,000 worth of food and stuff so we could stay in the house. I have pictures of myself going into Walmart with sunglasses, gloves, a mask. The managers at the store stopped me because they thought I was crazy. I was very paranoid. I was sanitizing labels so much that the ink on the wrappers was coming off.

But then the months went on and on, and I didn't die, and this virus didn't turn out how everyone said it would. Nobody I knew was getting sick. One day the government was telling us that the virus couldn't spread through the air, and the next day they changed their mind. One day it's "Don't bother to wear masks. They don't do anything." The next week it's "Wear a mask all the time. Right now. That's the new patriotism." It was like they were making stuff up, and then all the information started to get politicized. I started reading things about how the virus has a 99.5 percent survival rate, and how it's probably been here in the U.S. since last winter, so why are we shutting everything down and tanking the economy now in the months before a presidential election? When I see something that feels off, I have to question it. I don't rely on the mainstream narratives. I put the pieces together myself. Trump doesn't want to wear a mask, so now the media is holding that whole issue against him. Now it's mask, mask, mask. I told my wife: "Something's off here. I'm done going blindly along. This mask is just a propaganda tool. I don't care if there's conflict."

It's been worse than I thought. I've been in Starbucks, where you have a preteen working, and they tell me they won't serve me. I've been to Chase Bank, and the teller started telling me to put on a mask. She said: "My job is to keep you and others safe." I said: "No. Your job is to do my banking when I come in here. It's my body, my choice. Haven't you heard that before?" I've had people refuse to serve me and customers join in and start cursing at me—"Get the F out of here." I had one lady get in my face at the grocery store and scream at me, and it's hard not to get irate. I videotape everything. I stayed under control, but beneath the surface I was burning. I usually tell the employees to call corporate. I talk to them about the ADA exemption. I hand them a signed affidavit with my constitutional rights. I've had the police called on me many times. My wife had the cops called on her once at our local Petco when she was trying to buy cat litter, because they wanted her to mask up and she got loud and defiant. I have anxiety walking into these big box stores, knowing how many people are going to harass me. I don't even bring my son with me anymore. He's eight, and I can see the anxiety building in him every time we go out somewhere. He doesn't like being in conflict, but I tell him: It's our freedom. It's our rights. Some things are worth fighting for. No masks. No forced vaccines. Don't tell me what to do. Don't tread on me.

I put a group together here in Arizona. We're like-minded people, and we met at Trump rallies, and now we get together to go into stores around Phoenix. It's usually about twenty or thirty of us. We go in without masks and wave American flags. We tell the store clerks: "We will not comply." We're not there to hurt the store or damage property or make anyone feel bad. We sing and chant, and it's a really positive thing. It's "Free your face." "Stop hiding!" "Remove the diaper." "No more tyranny."

I go up to customers sometimes and try to start a polite conversation. I tell them: "Take off the mask. This is your country. Be free. Be proud."

Chapter 13

"I'm sorry, but it's a fantasy"

Jeff Gregorich, superintendent, on trying
to safely reopen his schools
Winkelman, Ariz.—July 2020

This is my choice, but I'm starting to wish that it wasn't. I don't feel qualified. I've been a superintendent for twenty years, so I guess I should be used to making decisions, but I keep getting lost in my head. I'll be in my office looking at a blank computer screen, and then all of a sudden I realize a whole hour's gone by. I'm worried. I'm worried about everything. Each possibility I come up with for this school year is a bad one.

The governor has told us we have to open our schools to students on August 17, or else we miss out on five percent of our annual funding. I run a high-needs district in middle-of-nowhere Arizona. We're ninety percent Hispanic and more than ninety percent free-and-reduced lunch. These kids need every state dollar we can get. But COVID is spreading all over this area and hitting my staff, and now it feels like there's a gun to my head. I already lost one teacher to this virus. Do I risk opening back up even if it's going to cost us more lives? Or do we run school remotely and end up depriving these kids?

This is your classic one-horse town. Picture John Wayne riding through cactuses and all that. I'm superintendent, high school principal and sometimes the basketball referee during recess. This is a skeleton staff, and we pay an average salary of about $40,000 a year. I've got nothing to cut. We've been spending money all summer to buy new programs for virtual learning and get hot spots and iPads for all our kids. If we lose five percent of our budget, that's going to be hundreds of thousands of dollars. Where's that going to come from? I might lose teaching positions or basic curriculum unless we somehow get up and running.

I've been in the building every day trying to get ready, sanitizing doors and measuring out space in classrooms. We still haven't received our order of Plexiglas barriers, so we're cutting up shower curtains and trying to make do with that. It's one obstacle after the next. Just last week, I found out we had another staff member who tested positive for the virus, so I went through the guidance from OSHA and the CDC and tried to figure out the protocols. I'm not an expert at any of this, but I did my best with the contact tracing. I called ten people on staff and told them they'd had a possible exposure. I arranged separate cars and got us all to the testing site. Some of my staff members were crying. They've seen what can happen, and they're coming to me with questions I can't always answer. "Does my whole family need to get tested?" "How long do I have to quarantine?" "What if this virus hits me hard like it did Mrs. Byrd?"

We got back two of those staff tests already—both positive. We're still waiting on eight more. That makes eleven percent of my staff that's gotten COVID, and we haven't had a single student in our buildings since March. Part of our facility is closed down for decontamination, but we don't have anyone left to decontaminate it unless I want to put on my hazmat suit and go in there. We've seen the impacts of this virus on our maintenance department, on transportation, on food service, on faculty. It's like this district is shutting down case by case. I don't understand how anyone could expect us to reopen the building this month in a way that feels safe. It's like they're telling us: "Okay. Summer's over. It doesn't matter what the virus is doing, we're tired of this. Time to get back to normal. It's been long enough." But since when has this virus operated on our schedule? Shouldn't we wait until it's safe?

I dream about going back to normal. I'd love to be open. My staff hates teaching these kids over the computer—that's not why anybody becomes a teacher. And these kids are hurting right now. I don't need a politician to tell me that. We only have three hundred students in this district, and they're like family to me. My wife is a teacher here, and we had four kids go through these schools. I know whose parents are laid off from the copper mine and who's going to the food bank because they don't have enough to eat. We delivered breakfast and lunches this summer to fight hunger, and we gave out more meals each day than we have students. The need is off the charts. I get phone calls from families dealing with poverty issues, depression, loneliness, boredom. We have several families that are worried about the possibility of suicide. Some of these kids are out in the wilderness right now, and school would be the best place for them. We all agree on that. They need to be learning. They need to be around their friends. But every time I start to play out what that looks like on August 17, I get sick to my stomach. More than a quarter of our students live with grandparents. These kids could very easily catch this virus, spread it, and bring it back home. It's not safe. There's no way it can be safe.

If you think anything else, I'm sorry, but it's a fantasy. Kids will get sick, or worse. Family members will die. Teachers will die.

Mrs. Byrd did everything right. She followed all the protocols. If there's such a thing as a safe, controlled environment inside a classroom during a pandemic, that was it. We had three teachers sharing a room so they could teach a virtual summer school. They were so careful. This was back in June, when cases here were starting to spike. The kids were at home, but the teachers wanted to be together in the classroom so they could team up on the new technology. I thought that was a good idea. I gave my approval. It's a big room. They could watch and learn from each other. Mrs. Byrd was a master teacher. She'd been here since 1982, and she had so many creative ideas. They delivered care packages to the elementary students so they could sprout beans for something hands-on at home, and then the teachers all took turns in front of the camera. All three of them wore masks. They checked their temperatures every day. They taught on their own devices and didn't share anything, not even a pencil.

At first, she thought it was a sinus infection. That's what the doctor told her, but it kept getting worse. I got a call that she'd been rushed to the hospital. Her oxygen was low, and they put her on a ventilator pretty much right away. The other two teachers started feeling sick the same weekend, so they went to get tested. They both had it bad for the next month. Mrs. Byrd's husband got it and was hospitalized. Her brother got it and passed away. Mrs. Byrd fought for a few weeks until she couldn't anymore.

I've gone over it in my head a thousand times. What precautions did we miss? What more could I have done? I don't have an answer. These were three responsible adults in an otherwise empty classroom, and they worked hard to protect each other. We still couldn't control it. We don't have control. That's what scares me.

We got the whole staff together for grief counseling. We did it virtually, over Zoom. There's sadness, anger, and it's also so much fear. My wife is one of our teachers in the primary grade, and she has asthma. She was explaining to me how every kid who sees her automatically gives her a hug. They arrive in the morning—hug. Leave for recess—hug. Ask to go to the bathroom—hug. Lunch—hug. Locker—hug. That's all day. Even if we do everything perfectly, germs are going to spread inside a school. I've been in this building for long enough that I don't think there's an illness left I haven't had. Strep throat. Pink eye. All of your common colds. It's inevitable. We share the same space. We share the same air.

A bunch of our teachers have told me they will put in for retirement if we open up the building this month. They're saying: "Please don't make us go back. This is crazy. We're putting the whole community at risk."

They're right. I agree with them one hundred percent. I

don't see how anybody could disagree. Teachers don't feel safe. We gave out a survey to all the parents, and most of them responded that they're "very concerned" about sending their kids back to school. So then why are we getting bullied into reopening? This district isn't ready to open. I can't have more people getting sick. Why are they threatening our funding? I keep waiting for someone higher up to take this decision out of my hands and come to their senses. I'm waiting for real leadership, but maybe it's not going to happen.

It's up to me. I can't bring students in here in good conscience. I'd rather sacrifice money over our health. It's the biggest decision of my career, and the one part I'm certain about is, it's going to hurt either way.

ALLISON WEBB, teacher in Canton, Ga.,
on becoming afraid of her students

I'm a teacher down into my bones. I've always been a teacher. It's my identity. It's who I am. I take my Spanish and French students on field trips in the summer, and I volunteer to supervise their Homecomings and Proms because I like to be with them. I love these kids. This is the only thing I've ever wanted to do, but right now the idea of walking into a classroom is giving me nightmares. It terrifies me.

Just a few days ago, I was in another Zoom meeting with our district about how we can minimize risk during the school day. President Trump says he wants all schools open for in-person learning, so our district has decided that's what we're going to do come hell or high water, but they still want us to limit our face-to-face exposure with students. Umm, what? How am I supposed to do that? I teach at a big public high school. I've got hallway duty in the morning, and that alone puts me in up-close encounters with maybe two hundred kids. None of them wear masks. None of them know how to social-distance. I've got thirty-four desks crammed together in my classroom for four classes a day, so let's call that another 130 potential exposures. Then I have my homeroom, lunch duty, recess, dismissal—I could go on and on. Each day, I come within six feet of at least four hundred students. How am I supposed to minimize the risks of that?

I've looked for any possible way to make myself feel better about this. I'd love to step into a classroom if I could do it safely. I'm so tired of looking at my computer screen I want to scream. I've had two of my own children graduate from this high school, and my daughter is in her senior year. This is my last chance to share a school with my child, and that's been special for me. I want this to work. When the district put out our reopening plan, I sat down right away and read it from cover to cover. I was excited that maybe they'd come up with some way to make it work, but no. There was no mask mandate for teachers or students. There was nothing about contact tracing, special cleaning, or reducing class sizes. It was one big whitewash. We're trying to teach a generation of students to be thoughtful and selfless, and this is the best we can do?

I barely slept that night. I went over my school day in my head and tried to make a list of the risks. Our kids share laptops back and forth, so how do I handle that? They sit at communal tables, and we can't do that anymore. They come up to the whiteboard and share the same markers. I'm a language teacher. We focus on pronunciation. I tell my students all the time: "Look at my mouth." The whole point of my classes is that we talk loudly in an enclosed space. You remember all those people who got sick at that choir practice out in Wash-

ington State? Eighty people caught the virus in an hour. I
have nightmares about that. We yell and laugh in my classes,
and we learn to roll our R's. You can't be silent in a language
class. These kids have to *see* me talk. My whole day is a super-
spreader event just waiting to happen.

I kept searching for workarounds. I went online to look
for a see-through mask so the kids could still see me talk. My
husband is an engineer, and he came up with an idea to build
a Plexiglas cage around me in the classroom. He sketched it
all out, and I had a covered desk and a corridor going down
the side of the classroom. It was like teaching from inside a
zoo. I showed the plans to the district, and they said no. They
thought it would give the kids anxiety. I'm sure that's true, but
what about me? What about my anxiety? The week before we
were supposed to go back to school, I could barely eat. I was
panicking. I was up all night. What am I going to do?

I made the decision to resign at the last possible minute. It
gutted me. I went into school to clean out my room. I pulled
twelve years of memories off my bulletin board: letters from
old students, pictures of our trips to Spain, postcards from
class pen pals in Europe. I prided myself on creating a global
classroom, and I've reflected a lot about how our culture has
managed this crisis in contrast with others. I have friends in
Spain who are going back to school because the caseload is
back down and they can be safe. I have friends teaching again
in Germany, where cases are down. There's a flaw in our culture in the United States, where somehow the independence
and individualism that we celebrate has cost us our ability to
sacrifice. It's "me, me, me." Unless it's convenient for us, we
don't really empathize with our fellow citizens and take care
of them. There's no capacity for sacrifice. We're selfish. I hate
to say that, but we are.

I still get all the all-staff emails from the district. They've been open a few weeks now, and it's more virus all the time. The football team had four cases the first week, so football was canceled. Dozens of students went into quarantine. Teachers got sick. A bunch more of them have decided to quit. It's a train wreck, and a few teachers have been like: "You saw this coming. This must be so validating."

But it's not. It's awful. I love this school and this community, and I'm broken up over this like everyone else. I'm grieving for us. I'm grieving for what we've become.

Chapter 14

"May rent. June rent. Late fees. Penalties."

Tusdae Barr, on being evicted from her
home during the pandemic
Houston—August 2020

There's no room for disbelief in my position. I've had other
people tell me: "Really? They're evicting you during a pandemic? That's crazy." I don't know. Maybe it is. I haven't had
time to sit and feel sorry for myself. First I was fighting in
court to keep the apartment, and then I was trying to get my
things out of there before the landlord padlocked the doors.

It was one big scramble. Where can I find a moving truck? How am I going to pay for storage? Who's going to watch the baby? Can we find a place to lay our heads, or are we going to become homeless? Which motels are safe and which ones are covered in virus?

It's a new problem every day, and that's how it's been for the last five months. I don't know what devil designed this maze, but congratulations. I've been running in circles and there's no way out.

The thing is, we were doing pretty great right up until March, when the virus hit Houston. Well—maybe not *great*. I've had struggles in my life. This isn't a fairy tale. But we were doing okay. We were making it. How about that? My daughter and my grandbaby were living with me, and my son and his family had their own place in the same apartment complex. I had a one-bedroom up on the fourth floor. We had air mattresses in there, but we made it up really nice. It was $900 a month and we were paying it on time. All of us had jobs. The whole family was together. Considering where I've been in my life, it was more than enough.

The layoffs started happening one after another when the city shut down. My son got sent home from his grocery store. My boyfriend did construction, and that dried up. I do office work for a temp agency, and they had nowhere left to send me. I called them up every day: "Hello? Please? I'll do anything you can get me." But eventually the phone calls stopped going through. It was just beep, beep, beep. I guess the temp agency went under and they must have closed.

We put together all of our savings to pay the rent in March. The federal stimulus money helped cover most of April. I started donating plasma every week for twenty-five bucks. I cleaned an old friend's apartment for twenty. I fixed up people's hair for whatever they could give me, but it was never enough. May rent. June rent. Late fees. Penalties. The hole kept getting bigger. Phone bills. Electric. Trash. Insurance. Water. We were stacking up bills and putting off our problems, and we knew it was only a matter of time. Houston had an eviction ban for those first few months, but that didn't mean they stopped charging us. The landlord kept knocking on our door, but all we had for him was more apologies. The court sent a letter in July telling us we needed to pay $3,000 or we would be evicted, and that's when everything started to spiral.

Stress has a way of bringing out the true version of people. I've had stress my whole life, so I guess you could say I'm used to it. I've lived through a lot of chaos. I had my son when I was fifteen, and my mom let me learn the hard way. She kicked me out and we bounced around the motels and rode the bus to hide from the elements. I had my tearful times and my rages, but then I decided to put it in God's hands. Now I pray and I let it go. I look hard for the small miracles that are coming my way. That's how I handle it. But my boyfriend, he started swallowing that stress and holding it down inside. He got angry. It brought out an ugliness in him I hadn't seen before, and he took all that anger out on me. He popped up behind me one day in the apartment and assaulted me. He bruised my wrists. He put a butcher knife to me. He chased me over to Walgreens, tackled me, and then attacked me in front of the store. The police came and arrested him, and he gave one of the police officers a black eye.

He's in jail now, and that's a good thing. He did what he did, and I don't want to give him excuses. There's no reason for violence like that. But there's also a part of me that knows where it was coming from. We were about a week away from getting evicted at that point. He still couldn't find work. My

unemployment payments were delayed. The county's emergency relief fund had run out of money. The court wasn't
budging.

I couldn't afford any kind of lawyer, so I went to court on
my own to appeal the eviction. They listened to my story and
gave me seven days to come up with $900. I think that was
their version of a good deal, being merciful, but I had nothing.
They might as well have asked me for a million dollars. How
do you come up with $900 when you don't know anybody
with a job? My son said we could come stay with him, but
with his kids living in that apartment, there wasn't any room.
My daughter is eighteen, and to me that's still a kid, and she
was worried about her baby. I told her: "It's all right. God
makes a way. We'll find somewhere to go." I printed up flyers
explaining our situation, and I walked around for a while to
the charities and the churches, handing them out. I came up
with a few hundred dollars, but the week went by too fast. I
got home one day and saw that the constable had put a new
sign on my front door. "You have 24 hours to vacate. Be out
by 9 a.m. tomorrow."

It helps me to focus on the good things. What are the good
things? The U-Haul place opened at seven in the morning,
and that was lucky, because some mornings they open at nine.
The nice lady at the counter listened to my story and rented
me the truck without asking for my driver's license, and thank
God, because I didn't have one. I needed help moving my furniture down from the fourth floor, and I just so happened to
see one of my son's friends sitting outside a convenience store,
and he agreed to help us move out even though I couldn't pay
him. It started to rain, and the constable isn't allowed to evict
in the rain or the snow, so that bought me a little more time
to get my things. We were able to move out the table and the

couch that my aunt blessed me with, and those are two of my most prized possessions. I was driving the U-Haul over by the cheap motels, trying to find a place that was clean enough and might cut us a deal, and I was starting to break down when this lady saw me crying. It was already getting dark at that point. She came over and asked how I was doing, and I laid my heart on her, and then she paid for two nights at the Residence Inn. Can you believe that kind of generosity to a stranger? And that place was so beautiful. It had a kitchen and a living room with an extra bed. I got in the shower and I just cried. I told my daughter: "Let this be the memory of today. Hold on to this gratitude."

I went back to the apartment early the next morning to get the rest of my stuff, but they'd already changed the locks. A lot of things were still left in there: my good pots, baby clothes, our vacuum cleaner. It's nothing we needed to survive. My uncle was a pastor, and he worked side by side with Martin Luther King, so a lot of times I go back to the story of Sodom and Gomorrah. "Don't look back. Don't turn into that pillar of salt. Keep on moving forward."

We bounced around the motels for a week, and now we're couch surfing. I'm at my aunt's place, and my daughter and her baby are with a friend. It's all very short term, and we're probably wearing out our welcomes, but there's nowhere else for me to go. My son sent me a message the other day to let me know that he's getting evicted now, too. He's a stocker at Walmart, but they cut back his hours during the pandemic. He told me: "Mom, it's impossible. I can't do this anymore. I give up." He hasn't answered his phone for a few days, and I'm getting worried about his mental health. He has a history of some darkness and depression. I'm not sure if he found a place to sleep or what. I keep calling him, but I don't know what to say. I'm his mother. I should be helping him somehow, but the truth is, I've got nothing to offer.

I'm trying to come up with a plan. I need to get us back together. I figure it's going to take at least $3,500 to rent another place, because landlords want double deposits and extra fees once you have that recent eviction on your record. I've been riding the bus and applying for jobs at the fast-food places around town. If you want to find work during COVID, you look for the drive-throughs. Those are the only places hiring, and I'm not too proud to take something like that. I've done Burger King, Popeyes, McDonald's, Chick-fil-A. I'd wear a chicken suit and dance in the street if they paid me $15 an hour.

It's been strange, going on all these interviews. I'm trying to present myself like I've got it all together, when things are falling apart. I force myself to act cheerful. I know how to put on that good face. I used to work for a nonprofit, and I'd get dressed up for these big functions and sit at a table with millionaires when we were trying to raise money. They would ask for my opinion or invite me to their homes for lunch, and I'd act like I was one of them. I'd look at their lifestyle and think: "Hey, one day that could be me." But I'm forty now, and the distance between that world and mine is still getting bigger. It's harder to put on that happy face. It's harder to pretend. The stock market is still going up and up, right? I guess all those people have already recovered. It's a different country up there at the top. Meanwhile, everybody I know around here is out of a job. Everybody is behind on the rent. Most of us are becoming homeless. The eviction courts are filling up. I'm worth exactly nothing on paper, so who's going to rent to me? I'm a second-chance case. I've got no home address, no employer, no car, no credit cards, nothing in savings.

I can go over a million reasons why, and all of them are true, but in the end what does it really matter? I've never found a landlord that likes to hear excuses. Sooner or later, the bill always comes due, and somebody has to pay.

JAYNE ROCCO, landlord in DeLand, Fla.

Let me guess: You're going to make me out to be one of the bad guys in this whole pandemic story, right? I've been a land-lord for twenty-five years, so I know how this goes. Yeah, yeah. I'm the villain in the story. I'm the monster. I'm the heartless, selfish, cruel woman who evicts innocent families and leaves them out to starve on the street. That's what you're thinking, right? I hear that stuff and worse from my tenants in an average week. Nobody cares about what the landlord is going through. Nobody knows my hardships. Nobody ever thinks about that.

Who do you think gets stuck with the bill every time one of my tenants doesn't pay? I do. I'm not some big corporation. I'm an old lady trying to save up to retire, and I'm getting ham-mered in this pandemic. I'm taking losses left and right. I'm falling behind on my mortgages. I can't even pay for my own groceries. I'm watching my retirement savings go up in flames, but you don't hear people crying about any of that. "She's just the landlord. She doesn't care. She's just the money lady."

I have ten properties in Daytona Beach. That's how I make my whole income. They're single-family homes—old places, lower middle class, nothing all that spectacular, but it's what I can afford. You don't come to me if you're looking for beach-front. I built my business from nothing ten years ago after my divorce. I was in a bad place, and my credit back then was terrible, but I worked my butt off and I found one lender who took a gamble on me. I bought one cheap little house. I rented it, flipped it a few years later, and then bought two houses with the profits. That turned into four, then six, and so on. Each one of my houses rents for maybe $1,000 or $1,200 a month, and I put that money straight toward the mortgage. A lot of landlords are like me—your scrappy, mom-and-pop-type operations. I might make $40,000 in profit during the whole year, but that's enough for me to live on. I'm not complaining. I was doing fine until this virus showed up.

It hit me in March along with everyone else. I know most of my tenants on a personal level, and I hear about their struggles. You see the messy side of people's lives when you're the land-lord. You get the tears and the excuses. I have one guy who works as a valet at a hotel, and the hotel had no guests, so his job was basically shut down. I can understand that. He called me up and told me he was going to be at least a few weeks behind. I was like: "Okay. Thanks for being straight up with me. I'll wait." Then I have another couple living in a fixer-upper, and both of them work at a restaurant, so they had nothing until they started receiving their unemployment. It was: "Please, Ms. Rocco, can you work with us?" Fine. Yes. Of course. I'm compassionate. I feel for these people. The econ-omy is falling apart and the situation is impossible. These aren't fake excuses. So what happens is, I'm giving them a break, but nobody is giving me that same leniency on the other side. I don't have the same option. I'm getting squeezed. If I don't pay the mortgages on these houses, I lose my whole business. So I kept reaching into my own pocket to pay the mortgage. I went through all of my savings in those first few months to cover about $5,000 in losses, and that was before things got really bad.

You always have problem tenants in this business, and lately I've had a few giving me nightmares. One guy, he was down-and-out when he first came to me, but he needed a place to live, and he was trying to get back on his feet, so I gave it a shot. Nice guy. Said all the right things. I have a soft heart, and that's not always a good thing as a landlord, because it turns out this guy was a drug addict, and the pandemic sent him back into addiction. He didn't pay me for March. He didn't pay April or May. I asked him to sign up for some free rental assistance with the county, but he never followed through on it. So now he's just using this pandemic as an excuse and taking advantage. I tried to offer him discounts and payment plans. I even offered to forgive all the debts and hand him $300 in cash if he would just agree to move out of the house, but he's not an idiot. He was living there for free, and that's better than someone giving you $300. Free rent is as good a deal as you can get. He started ignoring my calls.

I hate evicting people. It's the worst thing about this job, but sometimes you've got no choice. I filed for eviction on him and took it to the courts. But because of this pandemic, the government has all these federal moratoriums, state moratoriums against eviction. There's nothing you can do. Everybody wants to protect the renter, right? So my eviction was frozen, and I kept going thousands in debt. This guy keeps living in my house without paying one cent for all of June, July and August. I'd just rehabbed this house, and I would drive by sometimes and see new holes in the wall, graffiti on the side-walk, syringes on the steps. It was nasty. It was hard for me to sleep.

I talked to a lawyer. I launched a petition to try to get the governor to start a fund to help all of us landlords. Why are we the ones bearing the brunt of all this? We have living expenses, too. I called the police, and finally they arrested my tenant for possession of heroin and a bunch of other things. The police took him out of the house. That's the only way I was ever going to get it back. I went in and spent another $3,000 to fix it back up and get it back on the rental market.

My new tenant has been in there now for about five weeks. I didn't have time to properly vet her—I needed to get that initial rental deposit right away to pay off some of my debts. She's a nice woman, but she's already struggling, and she's already telling me she can't find steady work in this pandemic. She's been sending me some of the usual excuses, and in this economy they're probably true, but that still doesn't pay the rent. She had $1,200 due to me last weekend, and guess what? I'm still waiting.

Chapter 15

"How is this possible? What are the odds?"

The Graveson family, on the unpredictable
variance of COVID-19
Arlington, Va.—August 2020

MATTHEW GRAVESON, 16: Everybody keeps saying that what happened to us is a miracle, and I know that's true. But another part of me is like, "Really? You think I'm lucky?" Because I don't always feel lucky.

At every stage of what we went through, I look back and it's like: How is this possible? What are the odds?

TIMOTHY GRAVESON, 15: I didn't even know what a ventilator was before all of this. I didn't know about ECMO or life support. I never had any idea how bad this virus could get, which is probably the only thing that kept me from panicking.

GEORGE GRAVESON (FATHER): I got the virus first. We were wearing masks and taking aggressive measures to social-distance, but I must have picked it up at the grocery store or somewhere. I had a tiny cough, and then I had chills one afternoon. It was the standard version that millions and millions of people in this country have gone through. You get a bit sick and then you recover. It wasn't anything huge. It wasn't super drastic, and if one person in the family was set up to get hit hard by this virus based on age or whatever, it was probably going to be me. I have some of the preexisting conditions and markers. So when I came through it very quickly, it was like: Whew. That's behind us. It was a massive relief.

SHERRY GRAVESON (MOTHER): I tested positive and it was nothing. I barely had a sniffle. No fever. No cough. If I didn't get tested, I never would have known I was ever sick. Why? Why was it nothing? The doctors say they don't know why some people are asymptomatic and others are dying. Just about that range! It's beyond unpredictable. My daughter was in the house with us the entire time, sharing all our meals, and she never even got it. She tested negative. None of it makes any sense.

MATTHEW: We were so careful. I wasn't allowed to see my friends, even though I asked my parents like a thousand times a day. We wore masks. We quarantined more than any family I know. I remember texting one of my friends and saying: "Even if I get the virus, I'm not that worried. Maybe for my parents or my grandparents it is a big deal, but for us it will probably be like a cold." I thought kids weren't supposed to get the virus as bad. I thought it wasn't as common for kids to infect other people and pass it on.

SHERRY: Matthew started complaining about a runny nose on a Tuesday. Then it was pink eye symptoms on Wednesday, and the fever came after that, and it was very persistent. We tried every possible way to manage it at home, but his temperature wouldn't come down. He was lethargic and he said his whole body ached. We were calling the doctor and saying: "It wasn't like this for us. This is different. Why would he have something different if he caught it from us? Why isn't he getting better?" They were perplexed. They just kept saying: "There's a lot about this virus we still don't understand. We're asking the same questions."

GEORGE: This is a kid who's the example of young and healthy. He works out all the time. He's on all the sports teams at school, and now we're watching him struggle with his breathing to make it up the stairs. It was alarming. We took him to the doctor and had all kinds of consultations. We started to hear warning phrases from them about oxygen supply and multi-system failures. None of it added up in my mind. The decline was impossibly fast.

SHERRY: They told me to monitor his oxygen level. I checked it all the time. They told me it should be 98, 99—that's normal. I'd stand there and pray over him and wait for the number, and for the first days it was always pretty good. Then late one night, something told me to go check again, and it was dropping. First I checked and it was 91. Then I checked a minute later and it was in the 80s. It became that fear of the next number—not knowing, and not having any control. I called 911. I rode with him in the back of the ambulance to Inova Fairfax, near where we live in Virginia, and George stayed home to look after our two younger ones. The doctors did the X-ray, and it showed he had some pneumonia. They said they needed to take him upstairs so he could be admitted. They told me: "I'm sorry, but you have to go home. You can't be here as a visitor because of our protocols." That was their rule because I'd tested positive for COVID. They wanted to kick me out.

MATTHEW: That was the first time I felt afraid. I didn't want to be alone. I was still trying to put on a brave face, but I'd never been in that kind of situation before. Each time the nurses would walk away, I kept asking my mom if she would have to leave. She was saying: "No way. I won't. I can't. No."

SHERRY: I explained the situation to doctors, managers—I had like five different employees I was talking to. In no universe did I think I would actually have to leave. I said: "Are you kidding me? He's a minor. He's scared. This is my child and he's really sick." But they wouldn't budge. It was their rule. They took him upstairs, and I stood in the waiting room, and what was I going to do? I'm proud of myself for keeping calm. I got in the car. I felt sick. It was wrenching. I'll never forget that drive home. I went into the house and tried to hold it together and make sure the other kids were okay, but inside a part of me felt like it was almost dying.

TIMOTHY: After Matthew went into the hospital, the virus was in my head. My parents were trying everything to keep my sister and me safe. They would bring us food upstairs on a tray, set it outside my door, and then I'd wait until they got back downstairs before I even opened my door to get the food. I barely left my room for a week. I was just in there playing video games. I was still healthy at that point, but I felt like I was waiting to get it. I would blow my nose and be like: "Do I have it? Is this it?" Drinks would go down the wrong pipe, and I'd be like: "Oh no! Am I dying? Is it happening?" And then one day, I couldn't catch my breath, and I didn't have to guess anymore.

SHERRY: We took him to Inova—and it was the same thing for both boys. The doctors tried to help their lungs with supplemental oxygen, but it wasn't enough. They tried high-flow oxygen, flipping them over on their stomachs, different drugs, and finally a ventilator to breathe for them. They didn't have any explanation for why the virus was acting so aggressively. At each stage, they kept telling us: "Maybe this will be enough to support their recovery." But it was just down, down, down. Matthew's kidney started failing. He had liver problems. It was hour by hour whether they were holding through the night. At one point the doctors called and said they needed to rush Matthew to the other side of the hospital, because there was only one intervention left that might save his life. I wasn't there with him. I couldn't see him or hold his hand. The doctors were saying: "We need your approval to go ahead with this. It's crucial that we move very quickly." There was a witness on the phone, and it was: "Now. Now. Now."

It was all too much, and then a week or so later we went through the same thing again with Timothy.

TIMOTHY: All of a sudden it felt like fifty nurses and doctors came rushing into my hospital room. I couldn't control my breathing. They were like: "Just lie back and it will all be over." I kept thinking: I need to be asleep. Fall asleep. Please, fall asleep. I could hear a million beeps and so many footsteps. That's the last thing I remember.

ERIK OSBORN (DOCTOR): We used ECMO machines for both boys, and that's the highest form of life support we can offer. Ninety-five percent of our ECMO patients would be dead without it. The machine essentially replaces your heart and your lungs by pumping blood out of your body, oxygenating it, and then sending it back. It's a last resort. We're pushing the boundaries of physiology and bringing a patient to places they would never have gone on their own. You're suspending someone in the state right before death and keeping them in that place for days or weeks so their lungs have a chance to recover. It's become a crucial tool to fight this virus. We do our absolute best with cutting-edge medicine, but the odds are still long for any patient at that point. A lot of times nothing will work. That's the hard reality.

GEORGE: The boys were in rooms 618 and 620, and at that point we were allowed to visit for one hour a day. They were connected to these big bags of medicine, with tubes and wires running everywhere. They kept the room very cold. I remember going in there and almost shivering.

SHERRY: It was hard to get in the car and know what you were going to see. It was hard to go there, emotionally. I had to remind myself: It's still the boys. They're still alive. You're going to see the boys. We would rub their hair and hold on to their hands. We would read to them, sing, pray with them, and try to have a conversation. There was no reciprocation. We'd search for anything hopeful until the hour was up and we had to go back home.

GEORGE: Sometimes I would close my eyes and hold on to the fact that at least their hands felt warm.

SHERRY: I understand a lot more now about what faith is really about, which comes when you have nothing else left. I'm an organist at three churches in Alexandria, and we had daily prayer groups, people praying from over fifty countries, people sending us monetary support, food, favorite scriptures. I wrote down about a hundred verses and plastered them all over the house—on the microwave, doors, TVs, mirrors. I wanted reminders everywhere. "In this world you will have tribulations, but take heart. I have overcome the world."

DR. OSBORN: You get into thinking about that nebulous

line between life and death with these patients. For Matthew, I remember days when we would bring down his medication and try to wake him up to see how he was doing, and we got nothing back. He wasn't responding. You had no idea where his brain was, or if it was okay, or if he was still in there, or if he was recovering or reaching the end. We have the best medical advancements available to us, but at that point what you're doing is trying to stimulate the patient by grabbing their shoulder, looking into their eyes, and seeing if their eyes respond to you. This will sound unscientific, but there's something about looking into a patient's eyes and seeing that they're still a living being, and maybe they recognize they're looking at another living being, and you can see and feel their life force.

The nurses can sense that all the time, and I trust that every bit as much as anything I can get off the monitors. "Did he follow commands?" "No." "Did he move his legs?" "No, but he's in there. He's still in there."

MATTHEW: I remember dreams. I was in some abandoned hospital, and I was chained to the bed. I kept fighting the chains and trying to escape from the room, but I was trapped. When I woke up, I couldn't tell what was real and what was in my imagination. I saw a hospital room and a whiteboard on the wall, but the day and the month didn't make sense. I was like: Is this another dream? Where am I? What happened to the last month and a half?

TIMOTHY: When I woke up, I still thought I was dreaming. Everything was pitch-black. Sounds came back first. I heard beeping again, and I could feel the restraints on my arms. I couldn't move my arms, legs, chest, anything. I didn't believe I was back in reality until my mom started showing me videos of my teachers and everyone praying for us on Facebook. Then

I was like: Okay, wow. I guess this must be real. This is too crazy even for a dream.

SHERRY: All the pain and anguish of that month was nowhere near as intense as the joy when they came back. Matthew opened his eyes, and right away he tried to bust out of his bed. He still had his trach in his neck, but he saw me, and he tried to hug me. The nurses were crying and losing it. "He knows who I am!"

GEORGE: The relief that they recognized us, and reached out for us, and made the same jokes—it was euphoria.

SHERRY: The community arranged a parade when the boys came home about a month later. It was too much for them to walk, but we rode in a car behind a police escort. We had fire trucks and hundreds of people standing alongside the road holding signs and balloons. The boys were speechless.

MATTHEW: The support is mind-blowing, and I'm grateful. But there's also a part of me that's still like: Why me? I was healthy. I never really got sick. Even during the quarantine, I was working out and getting stronger, and then I ended up losing like fifty pounds and waking up as, like, a skinny, weak kid. I thought kids weren't even supposed to get this, and now I feel like a different person from it. There's a lot to rebuild.

TIMOTHY: Physical therapy, occupational therapy, speech therapy. We have so much of this stuff that it's like having a full-time job. It's a miracle to be alive, but it would have been luckier not to get it.

GEORGE: It's still a lot for us to process. For both boys to get sick like that? For both to get better?

SHERRY: They're back home. They're recovering. I go to bed at night knowing we're all here under the same roof. But I keep waiting for more answers. It's like we don't know why this happened or what might happen next. Can they get sick again? How long do they have antibodies? I asked every single doctor, and they said: "They probably have some kind of immunity, but we don't know how long, or how much, or what happens if they get it again. It's a novel virus. It behaves unpredictably."

Chapter 16

"Mom, help help help help!"

Jessica Santos-Rojo, on working,
teaching, and parenting at home
Gilbert, Ariz.—September 2020

Everybody's doing this, right? It's normal. It's basically what
parenting is now without schools. That's what I keep telling
myself whenever I'm hiding in bed for a few extra minutes and
trying to get myself ready to face the chaos of the day. It helps
me to think about all the other families in the same situation.
Otherwise, it can start to feel impossible, like this is all there is
and this is the only house left in the entire world.

We don't leave here anymore. We're trapped. We're on our
own island, and there's no way out. Every school lesson for the
kids, every bit of work I can somehow find time for, every meal,

every chore, every activity, every thought or conversation—it all happens inside the walls of this house. How does anyone live like this without going totally crazy? My preschooler has his little desk out in the hallway where he does video lessons and sings his school songs. My fifth-grader, third-grader and kindergartner are set up with computers for virtual learning in the living room. The kitchen has been taken over by storage bins for school supplies. The dining area is for virtual P.E. I basically roam from desk to desk to help with their log-ins and passwords and all the other problems that come up. "Mama, I need help. Mom!" I'm trying to do my job so I can pay all the bills. I'm trying to take college classes online at night to get myself into nursing school. How? When? Sometimes I go into the bathroom for a few seconds so I can take a breath or send an email and lock the door.

This is our fourth or fifth week of virtual learning in Arizona, even though it feels like we've been living like this now for five years. I guess you could say the whole experience has been humbling. My kids are remote learning for at least the first quarter, and then the school is going to reevaluate. As much as I hate the situation, it's the right decision. The virus numbers are still bad here. When the school made it official, I was like: Okay. I'm a single mother. I'm used to juggling everything for my kids. I can handle this. I manage a doctor's office, and they told me I could work from home. The school emailed advice on how to prepare our house for distance learning, and I took it up to the next level. I printed daily schedules for each of the kids and posted it to their desks. I got them color-coded bins for all their supplies. I created a whole system of "Mama Bucks" as a reward to keep them motivated, so they could earn fake dollars for movie nights and treats if they had good behavior. I hung a poster on the kitchen wall with our class rules. "Be respectful." "Raise your hand before talking." "Do your best." "Work hard and have fun."

They're good kids. They tried their best to follow all the rules, but they get sick of staring at a screen all day and they want to be with their friends. My oldest two boys deal with ADHD, and how many four-year-olds are going to be totally self-sufficient? At the beginning, I thought I might get an hour or so to myself while they did their work, but they need me every two minutes. It's relentless. I was staying up most of the night to get my own work done when I should have been sleeping. I was trying to squeeze in a few minutes here or there to answer urgent emails during the school day, but I guess that was still too much. I started getting emails from their teachers: "Why is this assignment missing?" "Why wasn't he signed in to the video chat for math?" "Is there any way to offer them more home support?"

I cut back my hours to part-time at the end of the first week, and lately my hours have been going down even more. I'll be lucky if I can find ten hours to work this week, and that's a problem, because we could use the money. My boyfriend and I are spending through all of our savings. I barely make enough now to cover my car and my phone, but what choice do I have? The kids come first. They have to come first. My boyfriend pitches in with them a lot and my mom helps whenever she can, but they need me. They need every piece of me at every minute.

Their password isn't working to sign in. Their computer is out of batteries. Their video feed went dead. "Mom, what's six times nine? What's fifty-eight minus twelve?" They need me to give them another writing prompt. They want lunch. They want snacks. They want to quit doing their schoolwork and watch TV. They can't find a pencil. The pencil is broken. The pencil is too sharp. The eraser is the wrong color. They're bored. They're tired. They're hungry. They're whining. They're fighting with me, and they're fighting with each other. "Mom, why is he looking at me? Mom. MOM! Tell him to stop looking at me!"

Sometimes I'll be talking on the phone to an insurance rep for work, but it's an interruption every few minutes. The preschooler needs to nap, but he's overtired, so instead he's throwing a tantrum. He wants to go outside, but it's 110 degrees, so it's Disney Channel on a loop, and now I feel bad about that. My kindergartner is learning how to write his numbers and letters, so he needs everything. "Mama, help. Can you show me?" My fifth-grader is supposed to be writing an essay about negative political propaganda. "Huh? Mom, what do they even mean?" My third-grader wants help with his subtraction, so I'm trying to show him how to borrow the one, but it's: "No. No! My teacher didn't teach us that way." So then he's bouncing off the walls. He's fast-forwarding through all his school videos, getting up from his desk and telling me he's done with all his schoolwork for the day. "What? How? It's ten in the morning. How can you possibly be done?"

I catch myself snapping and losing my temper sometimes, and I hate that. I went through my own issues with being picked apart and bullied at school, and I still have the emotional scars from that whole experience. I was the Mexican kid, the fat kid, four-eyes. It created all kinds of self-worth issues that I carry to this day. I want so much better for my kids. I want them to know they're special and precious. I want them to feel unconditionally loved. I should be able to do that, right? I adore them. None of this is their fault. I can be patient. I can be calm and kind. How hard should that be? I hold up my hand and I tell them I need a minute to calm myself down. I walk away and try to get some air, but their voices carry.

"Mom, now! Mama, please! Mom, help help help help help!"

I can't find a safe place to explode. When this is finally over, I'm going to drive to one of those places where you can pay money to smash and destroy things, because I could have a lot of fun with a sledgehammer right now. But how am I supposed to let it out? There's never any time. There's never the right space. Maybe if I could get just one freaking minute alone, but how? I don't want to snap at my boyfriend, because he's doing everything he can and this whole thing has already taken its toll on our relationship. I tried talking to a few relatives, but they don't have kids, so they don't get it. They were like: "Is it *that* bad? Maybe you should go back to work." Oh, yeah. Like I wouldn't enjoy getting into my car and driving into the office right now, with my nice comfy chair and my personal coffeemaker and a door that I can close? Thank you. Thank you for that loving advice.

The best outlet I have is whenever my boyfriend watches the kids so I can go to the grocery store. I swear, I try to go there every day. Sometimes he tells me, "Why don't you do a big shop and get it all done at once to save yourself all these trips?" But it's like: "No. I *want* to go every day. I want to be someplace quiet without little hands tugging at me. It's for my sanity."

I swallow all of it. I keep it all down. I've learned how to hold back my emotions and internalize. My youngest was born at thirty weeks with major heart problems. He was teensy tiny, 2 pounds and 11 ounces, and he stayed in the NICU for seventy-seven days. I learned to compartmentalize. I would lose it in the waiting room all the time, but I never let him see me cry. I'd go into his room and sing to him, smile, tell him how proud I was. I showed him love and positivity, and believe that it helped. Isn't that parenting? You take on the hard stuff and you try to give them the good.

But lately, the exhaustion and depression has started to come out sideways. I'll catch myself in the shower when I'm finally alone, standing and crying under the water. I'll stay there for a few minutes and let it out. How many more days? How much longer can life continue like this? But then I hear them calling for me, and I put myself together and come back out to help.

Chapter 17

"The cure has been worse than the disease"

Mike Fratantuono, on the death of a family restaurant
Glen Burnie, Md.—September 2020

I grew up inside this restaurant, and now my wife's helping out in the dining room and my daughter's working at the hostess stand. This is a family place—four generations and counting. I know every inch of pipe and wire we have running through here. I've been the plumber, the busboy, the handyman, the butcher, the bartender, the prep cook, and the manager. I've done every job there is in this restaurant, and now I'm the one who has to shut us down.

It kills me. We were supposed to be getting ready to celebrate our sixtieth anniversary this year, and instead we're packing up and closing at the end of this month as another victim of this whole economic shutdown. I'll find another job, but it's more than that. This place is home. These walls are like a family photo album. I try not to get too sentimental about it, because it won't change a damn thing, but sometimes the stress hits me and my heart starts going like crazy. I get frustrated. It makes me angry. How does this whole situation make any sense? I don't know a single person that's gotten hit really hard by COVID, and I know that's probably lucky, but right now I can tell you about at least a dozen businesses going under and a few hundred people going broke.

I know this virus is real, okay? It's real and it's awful. I'm not into conspiracy theories or any of that. I'm not disputing the facts of what this virus can do. But as bad as it's been, our national hysteria is making this whole thing way worse. We allowed the virus to take over our mindset, our economy, our small businesses, our schools, our social lives, our whole quality of life. We surrendered, and now everything's infected.

I like to believe this restaurant is resilient. My father worked here at least five days a week right up until he died at age eighty-two, so that's in our DNA. We've survived fires, chef changes, and every other problem you can think of in this business. We saw this pandemic coming, and we hunkered down and decided we were going to outlast it. When the governor first shut us down in March for in-person dining, I got together with my brother and my cousin, and we agreed to think of this as an opportunity. The restaurant was doing pretty good at that point. We had a little money in the bank to spend. We said: "Let's reinvest it back in the business like it's supposed to be done." We remodeled the entire bar. We put in new bathrooms and new draft lines. The place was sparkling. We gave a tour to a few of our loyal customers, and they said: "Wow. When this place opens up, you're going to fill it every day. You'll be back bigger than ever."

At first, the state was only allowing us to do carryout. So, okay. You do the numbers. We've got five dining rooms, eighty-five employees, and 13,000 square feet of space, and now you want us to operate like your regular old burger shack? How many people do you know who are willing to spend $40 on a lukewarm steak that's traveled halfway across town? That's not our typical business. We're old school. We don't have frozen hot wings that we heat up and toss in a cardboard box in five minutes. We buy the chicken fresh, cut off the fat, pound it out and bread it. We take pride in the quality of what we do. Your wings take twenty-five minutes while you're enjoying a conversation and a cocktail or two with some family and friends. Dining here is an occasion. It's a social experience. But that's the old world, right? What choice did we have? We redid our carryout menu to cut the prices down and teamed up with DoorDash and Grubhub and every other place like that. We created themed menus for Mother's Day and Easter. We delivered charity meals out to the hospitals and opened up a crab trailer out front so we could sell by the bushel. We were spending $800 a week on carryout containers, and there's no real profit in any of it. You lose out on booze. You don't get the same tips. We kept hanging on through March, April and May, but our revenue was down by more than eighty percent. We had to lay off seventy-five people. I must have gained twenty pounds that month. I'm the one who has to call each one of them. I'm the one who listens to them cussing, crying, worrying about not getting a paycheck, begging me for a job. That's seventy-five families dealing with unemployment and financial hardships, and not because a virus made any of them sick.

I know, I know. People are dying, right? Why should anyone care about my little problems? I've had this conversation a hundred times, and I get it. But what I'm trying to say is, this whole shutdown has its fair share of victims, too.

When the state finally opened back up for outdoor dining in the summer, I put all my hopes into that. I was desperate. I ran out to every Costco in the area and bought any picnic table I could find. We got twenty of them for $150 apiece, so that's a pretty big investment. We roped off the parking lot and put out buckets of cold beer. We advertised it on Facebook and everywhere else. We tried to make it look nice, but it's summertime in Baltimore, and it's ninety-some degrees outside and you're selling cream of crab soup to people who are sweating out there on the asphalt. Nobody really came. Why would they come? We stood there for nine hours each day and we waited.

We got more creative. We kept on trying. We opened up inside at fifty percent capacity as soon as the state said that we could, and we had live music, themed dinners, mystery nights and a meal with a psychic. We've reinvented this restaurant a dozen times, but none of it worked.

And what kind of support did we get? More rules. More restrictions. More regulations, mandates and curfews. We have to close every night now at ten p.m., because I guess maybe COVID comes out at 10:01—except apparently not at the casino down the road, because they have political leverage, and they get to stay open as long as they want. You can sit at our bar and have a drink, but you can't get up and mingle, because that's considered a health violation, so now I'm trying to serve you and police you at the same time. "Thanks for coming! It's wonderful to see you. We really need your business. Now put on your mask or sit the hell back down!" I have to buy disinfectants, individual ketchups, paper menus, and personal salt-and-pepper shakers or else I might get fined. All of that stuff is expensive. That's costing us more money that we're not making back when the restaurant can only be filled to fifty percent. My employees have to wear their masks all the time, even when they're alone in an office, but meanwhile we have a group of forty people in the dining room with no masks on, and that's deemed safe because they're sipping on glasses of water? None of it makes any sense.

We had a customer come in the other day, and she couldn't get seated because we had a few other tables, and we'd hit our fifty percent capacity in the one dining room she likes. I had to ask her to wait. I'd rather sit her down and take her money. I hate this rule, but now I have to enforce it. But she could see all the empty tables, and she started pointing them out to me and asking why she couldn't go sit. I tried to explain, but she didn't understand it. She got mad and decided to call 911 and tell them we were over capacity. Two armed police officers came through here. The whole foyer was empty. There was nobody at the bar. I've got a max fire rating on the wall for 323 people, and they couldn't count up to 100. I told them: "I dream about being over capacity, but I doubt it will ever happen again."

It's like Trump said: The cure has been worse than the disease. Nobody has gotten sick in this restaurant. My staff is all healthy. If you ask me, people spent too much time at home watching the news all day, drinking in this hysteria until they were spraying down their groceries and afraid to leave home.

It became another anti–Trump thing in the press. They won't leave this guy alone. The impeachment didn't work, the killer bees didn't work, so let's blow COVID out of proportion and see if it hurts him. But it's the rest of us that got hurt. It was day after day of failure. It was a slow and painful death.

We went to see our accountant at the end of the summer, when our reserves had gotten really low. He was a friend of my dad's, and he's been with us a long time. He's seen this business grow over the decades. I trust him. He looked over the numbers but he didn't say much, and that's not like him. I said: "Give it to me. What would you do?" He said the way things were going, we'd have nothing left to lose within a few months. He told us not to take it too personal. He said the big national chains might be able to survive this pandemic for a year or two, but all the family businesses that he works with are closing right and left.

We made the decision right then. There wasn't much to discuss.

Our last day is September 30, and then we're done.

Chapter 18

"All the way down the rabbit hole"

Tony Green, on dismissing, denying,
contracting and spreading COVID-19
Dallas—October 2020

I remember the first thing I thought when I woke up in the
hospital: The coronavirus is real. It's real and it's awful. How
embarrassing is that? I'm ashamed to talk about all of this. I
know what it's like to be humiliated by this virus. I used to
call it a "scamdemic." I thought it was an overblown media
hoax. I made fun of people for wearing masks. I called them

"sheep." I went all the way down the rabbit hole and then fell hard on my own sword, so if you want to hate me or blame me, that's fine. I'm doing plenty of that myself.

The party was my idea. That's what I can't get over. Well, I mean, it wasn't even a party—more like a get-together. There were just six of us, okay? My parents, my partner, and my partner's parents. We'd been locked down for months at that point in Texas, and the governor had just come out and said small gatherings were probably okay. We're a close family, and we hadn't been together in forever. It was finally summer. I thought the worst was behind us. We weren't violating any of the mandates or protocols. I was like: Hell, let's get on with our lives. What are we so afraid of?

Some people in my family didn't necessarily share all of my views, but I pushed it. I've always been out front with my opinions. I'm gay and I'm conservative, so either way I'm used to going against the grain. I stopped trusting the media for my information when it went hard against Trump in 2016. I got rid of my cable. The news has become just a whole bunch of opinion anyway, so I'd rather come up with my own. I find a little bit of truth here and a little bit there, and I pile it all together to see what it makes. I have about four thousand people in my personal network, and not one of them had gotten incredibly sick. Not one. Maybe a few people had it and then they recovered in a week, so that's sounding a lot like the flu. You hear the president saying that it's all overblown, and maybe some people can wear masks, but he's not going to bother with it. You start to hear jokes about, you know, a skydiver jumps out of a plane without a parachute and dies of COVID-19. A lot of people on conservative websites were saying all these numbers had been inflated. You start to think: Something's really fishy here. You start dismissing and denying.

I told my family: "Come on. Enough already. We're wasting away. Let's get together and enjoy life for once."

They all came for the weekend. We agreed not to do any of the distancing or worry much about it. I mean, I haven't seen my mother in months, and I'm not supposed to go up and hug her? Come on. That's not the way my family works. We have a two-story house, so there was room for us to all stay here together. We all came of our own free will. It felt like something we needed. It had been months of doing nothing, feeling nothing, seeing no one, worrying about finances with this whole shutdown. My partner had been sent home from his work. I'd been at the finish line of raising $3.5 million for a new project, and that all evaporated overnight. I'd been feeling depressed and angry, and then it was like: Okay! I can breathe. We cooked nice meals. We drank a little wine. We watched a few movies. I played a few songs on my baby grand piano. We drove to a lake about sixty miles outside of Dallas and sat out in the sunshine and talked and talked. It was nothing all that special. It was normal. It was great.

I woke up Sunday morning feeling a little iffy. I have a lot of issues with sleeping, and I thought that's probably what it was. Maybe insomnia. Maybe the wine. I let everyone know: "I don't feel right, but I'm guessing it might be exhaustion." I was kind of achy. There was a weird vibration inside. I had a bug-eyed feeling.

A few hours later, my partner was feeling a little bad, too. Then it was my parents. Then my father-in-law got sick the next day, after he'd already left and gone to Austin to witness the birth of his first grandchild. I have no idea which one of us brought the virus into the house, but all six of us left with it. It kept spreading from there.

I told myself it wouldn't be that bad. "It's the flu. It's basically just the flu." I didn't have the horrible cough you keep hearing about. My breathing never got too terrible. My fever peaked for like one day at 100.5, which is nothing—barely worth mentioning. "All right. I got this. See? It was nothing. What's everyone making this big deal about?" But then some of the other symptoms started to get wild. I was sweating profusely. I would wake up in a pool of sweat. I had this tingling feeling all over my body, this radiating kind of pain. Do you remember those old space heaters that you'd plug in, and the red lines would light up and glow? I felt like that was happening inside of my bones. I was burning from the inside out. I was buzzing. I was dizzy. I couldn't even turn my head around to look at the TV. I felt like my eyeballs were in a fishbowl, just bopping around. I rubbed IcyHot all over my head. It was nonstop headaches and sweating for probably about a week— and then it just went away. I got some of my energy back. I had a few really good days. I started working on projects around the house. I was thinking: Okay. That's it. Pretty bad, but not so terrible. I beat it. I managed it. Nothing worth shutting down the entire world over. Then one day, I was walking up the stairs, and all of a sudden about halfway up, I couldn't breathe. It was like a panic attack. I screamed and fell flat on my face. I blacked out. I woke up a while later in the ER, and ten doctors were standing around me in a circle. I was lying on the table after going through a CT scan. The doctors told me the virus had attacked my nervous system. They'd given me some medications that stopped me from having a massive stroke. They said I was minutes away.

I stayed in the hospital for three days, and I kept trying to get my mind around it. It was guilt, embarrassment, shame. It felt like I owed the whole world an apology for making light of this thing. I thought: Okay. Maybe now I've paid for my mistake. I got what I deserved. But the whole situation just kept getting worse.

Six infections in our family turned into nine. Nine went up to fourteen. It spread from one family member to the next, and it was like each person caught a different strain. My mother-in-law got it and never had any real symptoms. My father is seventy-eight, and he went to get checked out at the hospital, but for whatever reason, he seemed to recover really fast. My father-in-law nearly died in his living room and then ended up in the same hospital as me on the exact same day. His mother was in the room right next to him because she was having trouble breathing. They were lying there on both sides of the wall, fighting the same virus, and neither of them ever knew the other one was there. She died after a few weeks. On the day of her funeral, five more family members tested positive.

My father-in-law's probably my best friend. It's an unconventional relationship. He's fifty-two, only nine years older than me, and we hit it off right away. He runs a construction company, and I would tag along on his jobs and ride with him around Dallas. I've been through a lot in my life—from food stamps to Ferraris and then back again—so I could tell a good story and make him laugh. He builds these 20,000-square-foot custom homes for people, but he'd been renting his whole life. We decided to go in together on ten acres outside of Dallas, and he was finally getting ready to build his own dream house. We'd already done the plumbing and gotten streets built on the property. We'd planted fifty pecans and oaks to give the property some shade. He had his blueprints all drawn up. It was the first thing in his life that he had truly done for himself. He was over the moon. It was all he wanted to talk about.

He was on supplemental oxygen, but the doctors kept reducing the amount he was getting. They thought he was getting better. He was still making jokes, so I wasn't all that worried. He told me: "They've got you upstairs in the Cadillac rooms because you're white, but all of us Mexicans are still down here in the ER." I got sent home, and I had a lot of guilt about leaving him there. I called him at the hospital, and I was like: "I'm going to come bust you out *Mission Impossible* style." He said he preferred El Chapo style. We were talking about digging out a tunnel for him, and we were laughing so hard. Nobody made me laugh like that. I hung up, and a few hours later I got a call from my mother-in-law. She was hysterical. She could barely speak. She said one of his lungs had collapsed and the other was filling with fluid. The hospital put him on a ventilator, and he lay there on life support for six or seven weeks. There was never any goodbye. He was just gone. It's like the world swallowed him up. We could only have ten people at the funeral, and I didn't make that list.

I break down sometimes, but mostly I'm empty. Am I glad to be alive? I don't know. I don't know how to answer that.

There's no relief. I don't think I can live here in this house anymore, because it feels like it's cursed. This is the place it all started, and we've torn up this family with blame and regret. I can't escape this virus. It's all over my Facebook. It's the presidential election. It's Trump. It's the economy. It's what I keep thinking about. How many people in this family would have gotten sick if I'd never hosted that weekend? One? Maybe two? The grief comes in waves, but that guilt just sits.

Chapter 19

"Do something. Do something!*"*

Mary Jo Copeland, on choosing
decency over anxiety and fear
Minneapolis—October 2020

There's always a line. The line keeps getting longer. I wake up
at four in the morning to start helping these families, but this
pandemic never rests. I've been doing this work for forty years,
and I've never seen pain like there is right now. People come
here from all over Minnesota because they've lost their jobs,
their homes, their life savings—their dignity. They're carry-

ing around the hurt of what's been done to them. They've got nothing but anger, sadness and fear.

I had a lady show up the other day, another first-timer. Her life was falling apart in a hundred ways, and she started going on to me about this virus and all the unrest we have happening downtown. You have people protesting for Black Lives Matter and others marching around and talking about President Trump. This whole country is so divided. She was obsessing over this presidential election. She said: "I'm terrified right now about the future of this country and where we're going. It feels like I'm watching the whole world come unglued."

I told her: "Okay, then stop watching. What's something you can *do*?"

I've always tried to think like that. I'm not saying I don't have my own anxieties. I'm seventy-eight, and this virus has already set me back in a lot of ways. I've lost more this year than I ever have before, but what good has negativity and fear ever done for people? Nothing. Zero. You can waste your whole life as one endless complaint. You can worry your life away on the things you don't control. Okay, yes, this country has big problems. But who do you think is going to solve them? It's up to us. I believe in perpetual motion. Do something. *Do something!* If you see something that needs to be changed, try changing it. If you see somebody who needs help, help them. People act like that's saintly, but shouldn't it be basic? Why isn't it basic?

We're a one-stop shop to help the poor. We try to give people whatever they need: food, clothes, furniture, dental, housing assistance, money to pay their bills. We've been open every day since this virus hit, but it seemed like the rest of the city pretty much closed up. I don't accept government funds, which means I'm free from some regulations. We were the only place left serving meals downtown at the beginning of all this. We had five or six hundred people lining up to eat, and what am I going to do? Stay at home because I'm afraid I might get sick? Send people away if they aren't wearing a mask? Come on. These people barely had the luxury to worry about a virus. They were jobless. They were homeless. They had nothing to eat, and they weren't getting their food stamps because the county had shut down. I promised them: "I will not close." We served something like eight thousand meals that first week, and it's gone on from there.

Some days, there are two hundred people waiting to see me by the time I get in to work. Each one has an emergency. I open the doors and greet everyone as they come in. I ask their names and listen to their stories. A guy got laid off overnight at Mall of America, so now he's out there with four kids waiting for something to eat. A Somali family is about to get evicted from their apartment, and they're asking me to pay their landlord or else they have to move into a tent. We have a shelter across the street that houses six hundred people, and more than four hundred of those are kids. Every room is full. We call it transitional housing, but where are they supposed to go next? There's no hope for people in this pandemic. The safety net has been obliterated. Anybody who was vulnerable before this is now in a place where they are just hanging on to existence. We're going to be solving the problems from this pandemic for years and years.

I do walk-throughs of all the rooms in our shelter, and the other day I went to check on a Hmong family, a single mom and her kids. She lost her job because her restaurant closed down. Then the day cares closed, so even if she could find another job, she can't leave her kids alone. They're all stuck in the apartment. They've got maybe six hundred square feet.

The kids are trying to learn English, but there's nobody left to teach them. No more libraries. No more schools. No more free breakfasts and lunches during the school day, so now food is an issue for her, too. These people fall into the cracks, and the cracks become canyons. So now I'm sitting with them, trying to communicate, trying to figure out what I can do to alleviate a little bit of their suffering. I played with the kids and did some of my silly dances. I gave them cereal, clothes and a few dollars. It was nothing. It was nowhere near sufficient. I could have sat in that room and spent a year trying to help, and it still might not be enough. But the whole point is to try. The outcome might not always be in our control, but we are responsible for the effort. You can still make people feel seen. You can listen to them. You wouldn't believe the power that a little kindness can have on people. It restores our humanity. These kids were so grateful. They were looking at these little cereal boxes and just glowing. They had so much joy. I walked out into the hallway and I almost started to cry.

I know about sadness, and I still have a lot of that inside. I come from a hard life, and maybe that's why I was drawn into this. My father was a mean man. He'd tell me: "You'll never amount to a hill of beans, Mary Jo." He'd beat my mom all night and then demand his breakfast. She'd stand there bleeding, cooking his eggs. I went into a foster home for a while. I got depressed, addicted. I was thirty-eight years old when I finally decided to start volunteering and doing something with my life. All of our kids were in school at that point, and I had nothing to do. I was aimless. I felt lost in my own fear and sadness. My husband, Dick, he helped me understand: "Get outside of yourself. Make the world better because you're in it." I didn't have a driver's license. I'd never driven downtown.

I got lost on my way to Catholic Charities, but I knew God was calling me.

See, I'm nobody that special. I don't have much education. I don't have big money. I did what any person on this planet can do: I just decided to start.

Dick was alongside me the whole time. He helped me build my organization, fund-raise and recruit volunteers. I prayed a lot. I worked as hard as I prayed. Dick got a little mad at me because I never would take a salary for myself, but eventually he let that be. Not many people have a marriage like that. Then the dementia hit him, and he was in a nursing home the last five years. I'd go up there most days to feed him and help him get dressed. They had forty-seven people die from the virus in that nursing home, and the staff started telling me I couldn't come to visit. I went anyway. I had to see him. They came into his room one day and yelled at me, and I raised my voice. We had a confrontation, and I knew I couldn't come back. I will always believe that Dick thought: If Mary can't come, I want out of here. People are dying of loneliness right now. How sad is that? He only made it a few more weeks.

We had a small service. I had three priests, and they sang, and I was able to say a nice goodbye. But the emptiness when it ended—I don't think anyone can imagine. We'd been together since we were fifteen. I felt sick. I couldn't go back to the house. I was lost. I didn't know where to go.

Some of the volunteers looked worried when they saw me come back in my funeral clothes to help them serve the lunch. They said it was unusual. They thought I should take a few days off and rest my mind at home. I told them: "This is what might make me feel a little better." We had people waiting in line, and they all had their needs. I went back to work.

Chapter 20

"What could possibly go wrong?"

Chris Anderson, a Florida election official, on
running a fair election in a pandemic
Sanford, Fla.—November 2020

I believe in the power of positive thinking. That's the only way
that I can do this job and stay sane. I'm supervising a presidential election, in Florida, in a potential swing county, with all
kinds of voter distrust, while we're also dealing with a global
pandemic. What could possibly go wrong, you know? I try to
laugh about it. It's a crazy situation, but that's the story of my

life. I embrace the challenges. I read leadership books at night, and then I start each morning with an optimistic mindset that carries me into the day.

The first thing I do is check my messages. I got in to work the other day and there were ninety-six voice mails or texts from voters in this county. This job is a little like being a referee. If everything's going well, most people don't notice an election supervisor. But the crowd gets after you if there's confusion and fear, and that's what we're living through now. I had one message from a lady with asthma who hadn't been outside in months, and she was terrified of going to vote: "What if this is how I catch the virus?" Another woman had seen a story on Facebook saying mobs of poll watchers might stand near the voting booths: "Do they at least have to wear masks?" I hear theories about how precincts will become super-spreader sites, or armed militias will take over precincts, or postal workers will steal mail-in votes in some big conspiracy to hijack the whole election. I have to be the myth buster. I call every person back, even if it takes me all day.

My wife is right in this with me, and she saw me on the phone the other night and said I looked stressed. I told her: "Multiply that feeling by the 328,000 voters in Seminole County, and that's about where I'm at."

I don't blame people for being on edge. This is a historic election, and the stakes are high. It's a charged environment. I try to put myself in their shoes. I'm a visual learner, so I took one of our rooms at work and staged it as a precinct. We ran through the whole voting process to assess the risks. First you have to check in on a tablet that probably five hundred other people have already handled on Election Day, so that's a potential exposure. Then you sign in using a little stylus, and it's covered in all kinds of germs. The voting booths are too close together. They don't have Plexiglas barriers. Then you pick up the pen. That's another exposure. You touch the secrecy shield. Exposure. It's one hazard after the next. We have eighty precincts in Seminole County, and every one needed a safety overhaul. I came up with a list of about a hundred issues we needed to solve.

Like those styluses. That seems like an easy fix, right? Just make them disposable. Just buy them in bulk so there's one for each voter. But the price tag on that was about $250,000, and our whole budget for the presidential election cycle is $3.9 million. I wouldn't be able to pay my poll workers. I'd have to close precincts. So we started experimenting with stuff in the office, and someone on my staff figured out that if you roll tinfoil around the stem of a Q-tip and dampen the tip with a sponge, that transfers the electricity to the touch screen, and it actually works out pretty well. We started driving to every Dollar Tree in the area and bought all the Q-tips we could find. People were hoarding toilet paper, and meanwhile I was filling shopping carts with Q-tips, five bucks for 1,500 of them. We cut them in half and rolled tinfoil around the stems. My kids made some. My wife got into it. We stayed up late making Q-tips and watching *Game of Thrones*. Now we have more than a hundred thousand ready to go.

I feel like MacGyver in this job. One problem solved, another ninety-nine to go. Our pens need to be cleaned between each voter, so I found a friend who works in emergency operations, and he gave us three hundred gallons of disinfectant. We tested a bunch of pens in the solution, and they all dried out except for this one cheap brand, so I bought those out of stock. But now I've got another problem, because I need airtight containers in each precinct where I can store these pens in the disinfectant. Otherwise, this stuff is going to be sloshing all over the place, spilling on the floor, maybe dripping on things. Now I'm losing sleep over this. Now I'm thinking about election security and how I can make this all work. Then one day, I'm having lunch and eating my pho noodle soup, and I look down at the container, and a light bulb went off. This is it! It was just barely tall enough to fit a pen, and the lid closed up real tight. I started going to the different pho places outside Orlando, testing out these plastic to-go containers. I'm the crazy guy in the parking lot, kicking these different pho soups around and dropping them on the sidewalk to see which ones break. "It's okay, folks. I'm fine. I'm just trying to run a presidential election."

That's democracy in 2020. That's me trying to protect your right to go vote.

I joke about it, but this whole process is sacred to me. I'm a Republican, but I tell people all the time that my job is about principles over politics. I love this country. My life goes a different way if I'm born anyplace else. I was surrounded by so much failure growing up. My mother wasn't around. My father was into drugs and alcohol, until eventually he got AIDS. I'd get home from school and find cans of Natural Ice beer lying around, because he used those for his crack pipes. He got so high one time he tried to stab me with a butcher's knife. We moved to a new apartment whenever the electricity and the water got turned off, and that happened a whole lot. I became obsessed with doing something different with my life. I joined the military and deployed to Afghanistan after 9/11. I came back and started rising up through law enforcement. I made the decision not to be a victim.

We're all free to make our own choices in life. That's the

gift of being in this country. Voting is one of those choices. It's a reflection of our values and a chance to improve our lives. It might sound corny, but it's worthy of protecting.

I'd never really thought about election integrity until this position became available and the governor decided to appoint me. I was getting interested in politics, but supervisor of elections? I guess a part of me wondered if that might be a little dry. But it's been the opposite. Voting is under threat. We've got COVID-19 making us vulnerable every time we leave the house. We've got voting methods being dissected all the way up to the White House. We've got third-party groups trying to confuse voters, pushing out all kinds of misinformation, sending scam mail about registering their dogs or their dead relatives. I started an election academy for voters to see behind the curtain, and now we do trainings over Zoom. We go over ballot design, absentee voting—all of that. You might think it sounds kind of boring, but demand was off the charts. We spent eight weeks dispelling rumors and rebuilding that trust. I told them: "Every vote is going to matter in this election. Every vote will be counted, no matter how you cast it."

We encourage them to vote by mail, because to me that's the safest way when you're facing a pandemic. You never have to leave the house. We've gotten 106,000 vote-by-mail requests, which is way more than we've ever had before. That's how I vote myself, and that's the way the president has voted before in Florida. People are filing lawsuits against it, and they have all kinds of conspiracy theories about mail-in voting, but I don't understand the controversy. It's safe and secure. The data makes that clear.

Or you can go vote in person. We redesigned all eighty of our precincts, and now our voting booths have Plexiglas safety shields, and they're spaced six feet apart. We created a new staff position in each precinct for disinfecting and COVID safety. We have our cleaning procedures down to a science. We want to make it simple and worry-free.

You just sign in with the Q-tip, grab a pen from the pho container, lean down under the Plexiglas and cast your vote for president of the United States.

YVONNE BLACK, in Howell, Mich.

You're telling me this is how we run a fair election in America? It's a sham. It's a steal. Our democracy is practically being held together by mail-in votes and duct tape at this point. I don't even care how it turns out. I'll never trust an election again.

The alarm bells started going off for me this spring when they began pushing the mail-in voting here in Michigan like it was the Holy Grail. I'm sorry, but I'm a handshake person. I believe in showing up, showing your ID, showing your face, and signing your own name. If you care enough to vote, then get off your butt, go to the polls and get in line. That's something solid. I can trust that. But then the secretary of state decided she was going to send out ballot requests to every resident so they could vote by mail, and suddenly the whole thing started to seem like a free-for-all. Why does the secretary of state get to change our process? What about Congress? What about laws? I saw things on Facebook about these ballot requests going out to people who were dead, to people's pets, to some of our illegal residents. I have a friend down the street who got an application for her son, and he moved away to Arizona twenty years ago. People move all the time, and you've got these ballot requests floating around everywhere. That's fair? That's secure? I got upset. I got extremely concerned.

I started reading about ballot fraud and ballot harvesting. If everybody starts voting from home, and it's totally clandestine, the Democrats in this state can get away with whatever they want. You heard President Trump talk about it for months. What if they throw in thousands of fake ballots, or double-count, or invent people's names? He and I were both seeing the same red flags. So, okay, sometimes you have to do the basic math: They want to get rid of Trump at all costs. They want people to vote from home because we all know that kind of voting favors the Democrats. And now on my TV screen every fifteen or twenty minutes, I'm seeing these ads about how COVID is so awful and so dangerous, and we need to stay home to save lives. The fix is in! I've known for a long time that they are trying to steal our country and turn us all into California, but I never thought I'd witness them trying to steal an election with my own eyes.

I filed a lawsuit here in Michigan to force the state to stop mailing out ballot requests, but the judge tossed it out. I'm not someone who just rolls over. I signed up to be a poll challenger on Election Day. I started a voter information group on Facebook.

People come on there to bitch and complain about what's happening to our country, but that's not what I'm looking for. We have this McDonald's mentality, where everything should be easy. We've gotten complacent and apathetic in this country. I tell people: "I'm not looking for more whiners. I'm looking for warriors."

We're in for a hot war. I'm going to fight like the country depends on it.

KRISTIN URQUIZA, in San Francisco

I woke up on Election Day and I was so, so angry. That's every day for me, to be honest. I start thinking about this virus, and I think about how our president handled it, and I think about what happened to my dad. Do you ever have emotions that scare you? I don't know how to explain it. I grew up in Arizona, and we have these huge summer storms that rush through with a ton of rain and lightning, and they basically trample the city. That's how I feel, like I'm one of these summer storms with superhuman powers to mow everything down. It's too much. It's going to kill me if it stays inside. I'm always trying to find some kind of positive outlet, and at least one thing I can do is go vote.

My dad always liked Donald Trump. He trusted and respected him. You could say that was a difference between us. From the very beginning of the pandemic, I had this front-row seat to two versions of America. I'm in San Francisco, where it's all about science and public health, and this was the first place to issue an emergency order. Meanwhile, my parents live in Arizona, where the governor decided to be in lockstep with the Trump administration: late to close down, super-early to reopen—just all kinds of minimizing. I'm an only child, so I'm really close to both of my parents, and I sprang into that role of protector. At first, they were definitely listening to me instead of Hannity or whatever. They wore masks. They stayed home. They knew the virus was real.

The script totally switched when Arizona reopened at the beginning of summer. The president and the governor were very aggressive about saying it was safe, you could resume your normal activities, you could go out to eat with friends, and my dad wanted to believe them. We have this big, Mexican family, and my dad was incredibly sociable. He typically liked to go out and see people like five times a week. My mom called me after the stay-at-home order was lifted and said: "I'm worried. You need to talk to your dad." He was going out to a commencement party. He was watching sports with his friends and doing karaoke one night at a bar. I freaked out. I didn't have the best reaction. I remember calling my dad and yelling at him: "What the hell do you think you're doing?"

He was nice. He said he appreciated my concern, but I know he thought I was worrying for no reason. The whole situation made it seem like I was the crazy person. He kept saying: "I don't want to argue with you, but your opinion is kind of the outlier here. Why is every business reopening? Why is the governor loosening the restrictions? Why would Trump be telling us it's so safe?"

My mom called June 11 to tell me that he'd woken up with a fever. You know the rest. It's the same old story: agony and more agony. He died alone in the ICU with a nurse holding his hand.

A few days before he ended up going on the ventilator, I was talking to him on the phone and trying to poke around on the subject of his actions, and what did he think of the virus now. I didn't want him to feel like I was attacking him. My dad was not a quiet guy. He always had a lot to say, and his response to me—he paused for a really long time. He said: "I feel like they lied to me. I feel betrayed."

After he died, I had these episodes of mania, staying up all night fueled by rage. I know COVID killed him, but it wasn't the only cause. I'll read what I put in his obituary: "His death is due to the carelessness of the politicians who continue to jeopardize the health of brown bodies, through a clear lack of leadership, refusal to acknowledge the severity of this crisis, and inability and unwillingness to give clear and decisive direction on how to minimize risk."

My dad loved politics. He taught me that elections are about consequences and holding leaders accountable. I've never put more of myself into a vote.

Chapter 21

"Election Day is over, and guess what?"

Tom Dean, a physician, on treating
the world's worst outbreak
Wessington Springs, S.Dak.—November 2020

Election Day is over, and guess what? The virus is still here.
It didn't just go away like the president said it would. We're
not rounding any corners. Nobody I know in South Dakota
stopped talking about it just because the voting is done. How
could we? It's right in our faces. It's spreading. It keeps getting
worse.

Look, I'm as tired of hearing about COVID as the president
seems to be. I'm so sick of this virus, but what else should I be
focused on, exactly? We're going to have a new president, and
that's important, but for the moment nothing has changed.

What's even going to be left of South Dakota in a few months when Joe Biden takes over? I'm one of three doctors in this county, and we're right in the middle of one of the worst outbreaks in the world. We have to do a little bit of everything in a rural community, and this virus follows me wherever I go. We test people at our clinic, and right now more than half of them end up being positive. We give them supplemental oxygen in our local hospital until they get critical, and then we have to transfer them to the ICU in Sioux Falls. I'm also a medical advisor for our nursing home, and it just had a big outbreak. Now they have eight or ten empty beds.

Jerauld County is an out-of-the-way place. We don't have a whole lot, but we're proud of what we have, so it pains me that we're becoming famous now for our statistics. One in every twenty people here has gotten sick in about the last month. Our death rate is the highest in the country, but it's more than that. These aren't anonymous cases. These are my patients, my friends, my neighbors, my family. I know every single one.

We got lucky early in this pandemic, and I think that made us complacent. It was China, Seattle, New York. There was some hope in rural America that this might stay more of a big-city problem. We have about two thousand people here spread out over five hundred square miles of cows and wheat, so social distancing came naturally to us. The governor decided to kind of pretend it away and keep everything going as usual. She essentially told everybody: It's not really going to hit us. I think people got a little deluded. You didn't see too many wearing masks. The school opened back up pretty much right on schedule. They started playing football. The stands filled up every Friday night. I was waiting for the other shoe to drop, but for a long time it really didn't happen. I bet we went three months at the clinic without seeing a single positive test.

How do you go from nothing to the worst outbreak in the country? I don't know. I don't have a good answer. We started to see a few little pockets of it, but the virus seemed fairly contained. There was a small outbreak at a café in town, and that led to seven or eight cases in our farming community, but then it seemed to go away. There were a few cases up north at the beef jerky plant, but they seemed to stay mostly within the workplace. A couple of our students came down with it. Then a few teachers. I went into the clinic one day this fall, and the phones were ringing off the hook. Coughs. Fevers. Chills. We're the only shop in town. There's nowhere else for people to go. We tested people as they pulled up in their cars. We have rapid tests, so you hope and pray for those fifteen minutes. I was worried about how fast this could spread through town. We have one grocery store, one bank, one pharmacy—it's all the same petri dish. I was worried about what would happen if COVID got into the nursing home, where both of my parents lived. I told myself: Maybe it's just an outbreak of the flu or something seasonal that's going around.

Positive. Positive. Positive. Positive. We had eleven new cases within about five hours.

I can't even get my head around how bad things have gotten right now in the Dakotas. In this whole area right now, about one in every hundred people has an active case of the virus. We're not built for this. Our population is older, poorer, more obese. You name an underlying condition for this virus, and we have it, plus we don't have the resources to fight back. Our health system is overtaxed. These rural communities don't have hospitals or enough doctors. We don't have a single medical specialist in this entire county. It's me. It's always me. What kind of sense does that make? I'm seventy-five. I've got my own underlying conditions. I'm supposed to be retired right now and focusing on my woodworking, but if I do that, we're down to just two doctors for miles and miles. What choice do I have?

I've been trying to convince people that this is an emergency, and we need to change our behavior. Once you get sick, there's just not a whole lot that can really be done. We can give you supportive care, but a lot of what we're doing is just crossing our fingers. The thing you can try to control is to avoid getting it. I've been practicing here for forty-two years, so I'd like to think people trust me, but there's a strong independent spirit that I love about South Dakota. My great-grandfather came here as a homesteader in 1882 when it was nothing but wide-open spaces. My grandfather and my father hung on to our family farm during the Great Depression. We have a lot of people like that—stubborn and tough. They burn their own wood to get through winter. They take care of their land. They don't want to be watched over or babysat or told what to do, and I can understand that, but I'd like to believe we're still capable of making a communal sacrifice and rising to a historic occasion. Stay home. Be reasonable. Wear a mask.

I started writing a column for the local newspaper every week, the *True Dakotan,* trying to sound the alarm. "Don't let your guard down." "Armies that underestimate the opponent usually lose." "The threat is real." "We are in the middle of it." "Act as if you have the virus." "We are in trouble."

But the numbers kept rising. Some of the nearby ICUs filled up. The school got it pretty bad and eventually decided to close. Then the nursing home finally had its first positive test, and that was the phone call I'd been dreading. Those folks kept the virus out for more than six months, and they tried so hard. The administrators immediately sectioned off one area of the building and set up their own COVID isolation unit, but it's a small facility, and it was already too late. The virus snuck in and went wild. Within a few days, most of the residents had tested positive. Two-thirds of the staff was out sick. They had these frail, confused, sick residents, and nobody could really take care of them. They tried to hire some temporary help, but some of the new nurses refused to treat any COVID patients. "What? That's everyone! What did you come here to do?"

I'm not claiming to be any kind of a hero. I knew I couldn't go into the nursing home, even though I felt a little bit ashamed about that. I'm slowing down and I'm clearly in that high-risk group. I'm a cancer survivor. The nursing home doesn't allow visitors, and I hadn't been over there to see my parents in six months. My wife said to me: "Don't you even think about doing something crazy." I talked to the staff on the phone every day to see how they were doing with the outbreak, and I did a fair number of video visits. It killed me. I looked over the camera and saw a bunch of people on oxygen, wheezing, pale. It was just a matter of time. Some of those people had been my patients for forty years. You feel so helpless. I mean, how am I supposed to care for these people? Some have dementia. Others aren't speaking. Even if I could be there at the bedside, this virus makes you powerless. It kind of does whatever it's going to do. I've mostly provided moral support, maybe a little company to help with their loneliness. One woman had been my patient for twenty years, and she got it and died practically the next day. Another one of my friends—I guess I can't call him a patient because he never liked going to the doctor—he seemed to be getting better and then a few weeks later he died. There was no good news. It was one bad call after the next. My parents both tested positive—boom, boom.

My mother had been steadily losing ground for a while before COVID, not eating, sleeping twenty-two hours a day.

It was her time, if I'm being honest, and she kind of drifted off. My dad was different. He was clinically stable. He could get confused, but he was comfortable and fairly active. When the virus hit, he went downhill in a hurry. It attacked his lungs. He needed oxygen. His body got stressed. It took four days from his diagnosis to his death.

We wanted to hold a funeral outdoors while the weather was still warm, but my brother had a conflict, and then my granddaughter got sick with the virus at school. She probably gave it to my son-in-law, because now he's got weakness, fatigue, and shortness of breath. My daughter's been okay so far, but you have to assume it's a matter of time. Anyway. The funeral will just have to wait.

A lot of people have suffered worse losses to this virus. I know that. I've seen it with my own eyes. My dad was over a hundred. My parents lived a good life, and they were at the end of their road. They saw the best of this country. They got married seventy-six years ago during World War II once they'd finally saved up enough of their sugar rations to bake a proper wedding cake. They loved telling that story. Everybody was sacrificing for the war. It was a national effort. They were proud to be a part of it. They put the importance of the whole before themselves. That was the American way. The country had bigger problems, and their wedding cake could wait.

How can we get back to that spirit of sacrifice? What happened to us? My hope now that this election is over is maybe we can take a break from tearing each other apart and come together for the common good. Maybe that's naïve, but I believe in the best parts of this country. It's either that or give up, right? What other option is there? The virus is still raging, and there's no magic solution coming. It doesn't just magically go away on its own. It's up to us to stop it.

Chapter 22

"This is how we treat each other?"

Amber Elliot, county health director, on
the high costs of doing her job
Farmington, Mo.—November 2020

I don't really know if I should be talking about all of this. It
makes me worried for my safety. I've had strange cars driving back and forth past my house. I get threatening messages
from people saying they're watching me. I have to check the
security cameras every time I leave my office to make sure the
parking lot is clear. One of these people followed my family

to the park and took pictures of my kids. How insane is that? I know it's my job to be out in front talking about the importance of public health—educating people, keeping them safe. Now it kind of scares me.

But people need to know what's going on right now. There's this basic denial of science and facts. It's happening all over the country, and it's not acceptable. I know we can do better. We have to do better.

I don't base our whole response to this pandemic on my own *opinion*. That's what makes the backlash so confusing. This job is nonpartisan. I'm not political in any way. I based my work off of facts and evidence-based science, and right now all the data in Missouri is scary bad. We only have about 70,000 people in St. Francois County, but we've had more than 900 new cases in the last few weeks. Our positivity rate is twenty-five percent and rising. The hospital is already at capacity. They've basically run out of staff. We can't keep up. We've already had more than 250,000 deaths in this country, which is insane to me. It's an uncontrolled spread. I have these moments when it feels like I'm a nurse at the bedside, and my patient is dying, and I'm trying every possible intervention to save them. More social distancing. More masks. More contact tracing. Warnings and more warnings. What else can we try? But in the end, it doesn't matter how much you do. Nothing will work, because it almost seems like the patient is resisting your help.

I get the same comments all the time over Facebook or email. "Oh, she's blowing it out of proportion." "She's a communist." "She's a bitch." "She's pushing her agenda."

Okay, fine. It's true. I do have an agenda. I want disease transmission to go down. I want to keep this community safe. I want fewer people to die. Why is that so controversial?

We weren't set up well to deal with this virus in Missouri.

We have the worst funding in the country for public health, and a lot of the things we've needed to fight the spread of COVID are things we should have had in place ten years ago. We don't have an emergency manager in our county. We don't have anyone to handle HR, public information, or IT, so that's all been me. We didn't get any extra funding from the state for our response to COVID until last month, and by then we were already way behind. I'm young and I'm motivated, and I took this job in January because public health is my absolute love. Call me naïve, but I'm a true believer. This job doesn't pay well, but would I rather be treating people who already have a disease or helping to prevent it? That's what we do. We help keep our community safe. We work to protect our residents from sickness and disease. At one point this summer, I worked ninety days straight trying to hold this virus at bay, and my whole staff was basically like that.

We hired ten contact tracers to track the spread of this virus starting in August, when we finally got the money, but the real problem we keep running into is our community cooperation. We call everyone that's had a positive test and say: "Hey, this is your local health department. We're trying to interrupt disease transmission, and we'd love your help." It's nothing new. It's basic public health. We do the same thing for measles, mumps, and tick-borne diseases, and I'd say ninety-nine percent of the time before COVID, people were receptive. They wanted to stop an outbreak, but now it's all politicized. Every time you get on the phone, you're hoping you don't get cussed at. Probably half of the people we call are skeptical or combative. They refuse to talk. They deny their own positive test results. They hang up on us. They say they're going to hire a lawyer. They give you fake people they've spent time with and fake numbers that they want you to call. They lie and tell you they're quarantining alone at home, but then in the background you can hear the beeping of a scanner going off at Walmart.

I've stayed up a lot of nights trying to understand where this whole disconnect comes from. I love living in this part of Missouri. I know in my heart these are good people, but it's like we're living on different planets. I have people in my own family who believe COVID is a conspiracy and our doctors are getting paid off to inflate the death count. I've tried doing press conferences and dozens of Facebook Live videos to talk about the real science. Even with all the other failures happening right now, the science is the one thing we should be celebrating: better treatments, nurses and doctors on the front lines, promising news about vaccines. But the more I talk about the facts, the more it seems to put a target on my back: "We're tracking your movements." "Don't do something you'll regret." "We'll protest at your house."

The police here have been really great. The elementary school says they're watching over my kids and they're on high alert. I have a security system now at my house. I locked down my email and took all my family photos off of Facebook, but you start wondering: Is this worth it? Could anything possibly be worth it?

And then it got worse this fall around the whole masking issue. Our hospital was filling up, and they asked if we could do more in terms of prevention and masking. We put out a press release. We went to businesses and did trainings. We kept encouraging people to mask up, but it wasn't working. Only about forty percent of people in our community were wearing masks, so the health board decided to push for a mask mandate. Of course I was in favor of the idea. Of course it is the scientific, smart thing to do. But at the same time, the politics around it were incredibly toxic. I kept thinking: Is this going to blow up my life?

We held a public meeting in the auditorium. I knew it was going to be a circus. People were talking online about showing up with guns and ammo. I was scared. I gave my kids an extra hug that night and said the things you never want to have to think about. I asked the city manager: "Are you requiring masks in this building for the event? Because this is a public health meeting, and that's important." They said yes. But, of course, the first person that walks in the door says: "I go to church here in this same building, and they don't make me wear a mask, so I'm not wearing one now." So that ended up being an ordeal, and they decided to allow him in anyway. I asked him: "Can you please, please, please social-distance?" He told me no. It wasn't: "I can't." It was: "Hell, no. I won't. You don't get to tell me what to do with my own body. You can't take away my rights." It all went downhill from there.

We had more than a hundred people show up, and most of them spoke in opposition of masks. We do get a lot of thank-yous and support for our work, but those aren't usually the loudest voices, so sometimes they get drowned out. Our medical providers were at the meeting in their white coats, and three of them stood up to speak on behalf of masks. They had charts and tried to share some of the data about how this virus is decimating our community. These are doctors and nurses who risk their lives to treat this virus. They are shouldering the burden of this, but the crowd wouldn't even let them talk. They booed. They yelled out threats. They were so disrespectful. I was trying to take notes for our board, and my hands started shaking. Why aren't you listening? Why are you shunning the science? Why do you refuse to hear from the people who actually know about this disease and how it spreads?

The board decided to go ahead with the mask mandate anyway, but part of the community revolted. I got probably twenty threats in my email that next week. I started having anxiety about leaving my house. We did a survey a few weeks later, and mask wearing had actually gone down by six percent after the mandate. We required it, and people became more likely to do the opposite. It's anarchy. How do you even make sense of that? This is rural Missouri. We like to believe we take good care of each other here. We pride ourselves on being a down-home community that sticks together, and now this is how we treat each other? This is who we are?

I don't go out in public much anymore. It's in to work and then right back home. I don't want to be recognized. I don't want my kids to see any of that hate. The one place where I had to draw the line was that my son plays baseball, and honestly, his games are the most normal I've felt all year. But then, a little while ago, one of these haters followed me to a game and took a photo of me with my daughter. We were outside in the grass and social-distanced, so we weren't wearing masks. The photo got posted all over social media, and it was the usual comments. "Bitch." "Communist." "Hypocrite." My daughter has started to have some anxiety. My son said to me: "Mom, why does everybody hate you?"

I went in to work the next day, and one of my nurses came to see me in my office. She'd just had one of those nasty interactions with a denier on the phone, and she said: "I'm struggling right now with my motivation. It feels like some of these people are beyond our help. I need one of your little pep talks."

I told her: "I'm sorry, but I just don't have it. I'm tired of this. I'm so, so exhausted."

I've been living with that steady hum of tension and fear for almost a year, and I just can't do it anymore. It's not like it's going to get better anytime soon. I'm already getting push-back and hate mail from this community about how they're not going to take this vaccine from the Deep State, so I'm on guard against all of that. I keep saying my family is my number one priority, so at some point I have to keep my kids safe. Who am I going to protect? My kids, or a part of this community that seems determined to do everything it can to spread this virus, no matter what I say?

I can't do it anymore. It's making me crazy. I decided to put in my notice earlier this month. My last day is this Friday.

I've already accepted another nursing job. I'm not abandoning the community. I'm going to keep fighting this pandemic, but I'd rather not say anything much more specific about where exactly I'll be working. I don't want that target on my back. I'm ready to be anonymous.

Chapter 23

"This is it. This is who I am."

Kaitlin Denis, on what happens when
COVID-19 won't go away
Chicago—December 2020

I've been sick for 287 days. I've had a fever off and on now for most of the last nine months. I used to go to bed thinking: *Tomorrow.* Tomorrow it will start turning around. Tomorrow I'll feel better. I don't do that anymore. I don't have any energy left to summon that kind of hope. I'm coming to terms with the fact that this virus is not something I'm getting over. This is it. This is who I am.

I wake up in the morning now and I brace myself. What's it going to be today? Long-haul COVID is like a whole grab bag of symptoms. You reach in, and you never know exactly what you're going to get. How about severe nausea with a touch of dizziness? Or what if we throw in a headache and a little joint pain? Some symptoms never go away, but others are weird and they come and go: extreme fatigue, ringing in my ears, sore ribs, trouble breathing, ear popping, numbness in my fingers, excessive mouth watering, dizziness, light-headedness, brain fog. My vocabulary has started to disappear. It's like one out of every hundred words now is just gone. Sometimes my memory loss is so bad that it's like I have amnesia. The other day I woke up and tried to get on my running clothes. In my head, I thought I was fine. I thought I was going to go for a jog on Lake Michigan and then head in to work, but as soon as I stood up my heart rate started spiking by like fifty beats a minute, and it was like: Oh yeah, I can't go anywhere. I don't have a job anymore. I'm on long-term disability. What am I thinking?

A lot of days I hardly get out of bed. The fatigue flattens me. It's like the day never starts or the night never ends. It's all the same. It's a black hole. It's like I'm trapped inside this body and waiting for the hours to pass.

I'm starting to get more and more withdrawn. The whole thing is embarrassing. I get the feeling some of my friends think I'm being overly dramatic. What am I supposed to tell them? How do I even start to explain this? I don't know what's happening, and neither do any doctors. They just want to put me on antidepressants or send me to counseling, because medically none of this makes any sense. I'm barely thirty years old.

I just got married. I was totally, insanely healthy. Ten years ago, I was playing Division I college soccer, and now I can't go to the grocery store unless I ride around in one of those carts. It's like: Really? *Really*? Can it really be *that* bad? It seems pathetic to people. It seems pathetic to me.

When I first got sick, COVID was barely anything in the United States. It was the middle of March, and Chicago hadn't started any kind of lockdown yet. Nobody was wearing masks. You couldn't get a test if you hadn't traveled outside of the country. I started having a headache and a sore throat, but I gutted through it. My friends were like: "You probably saw COVID on the news, so now you think you have it." I work in finance. It's the kind of place where you're either at work or you're on your deathbed. It's that Wall Street culture of hand-to-hand combat, and that suits me. My husband tells me I'm a killer. I go crazy each morning waiting in line behind people who don't know what they're ordering at Starbucks. I don't like to slow down. I'm extremely competitive. I sucked it up for a while until my fever starting spiking, and then my husband started having symptoms, too. I called the Northwestern COVID hotline. They told me to go to the ER, but the ER said they didn't have any tests and there wasn't much they could do anyway. They told me to assume that I had it. They gave me some headache medicine and sent me home.

It was a rough next week for both my husband and me: chest pain, body aches, migraines. We ordered tons of Gatorade and we didn't leave the apartment for three weeks. We'd argue about who had to get out of bed to take the dogs to our balcony. We got a steroid inhaler to help with the breathing and basically just laid in bed, and then after a few weeks my husband started to feel better. He was going for runs again. He said: "Come on. At least go for a walk with me. You're okay.

We can do this." I tried. I tried to fake my way through it even though I could only walk like one block. I went back to work after three weeks. I was dealing with so much fatigue that I couldn't focus. I've had concussions playing soccer, and it was that same familiar fogginess. There was so much pressure in my head that it felt like I was hanging upside down. I would lie on my bed and move my mouse every once in a while so it looked like I was active on my work computer screen. I was making a ton of silly mistakes. Sometimes, when I would place large trades over the phone, I would forget what I was doing in the middle of the call. My job is all about the markets, and I couldn't remember the day of the week. I was like: Okay. Something is seriously wrong. I'm not getting better.

I've seen more doctors in the last six months than I did in the first thirty years of my life. I have to be my own advocate, and it's exhausting. First you can't get an appointment, and then you can't get insurance to pay. Then most of the time the doctor says he doesn't know, or maybe he doesn't even believe you. And it's like: What's even the point? I went to an internist in Chicago, and he sent me back to the emergency room, so that didn't help anything. I finally got in to see a rheumatologist, who referred me to another rheumatologist, who sent me to another specialist out of state, and then she sent me to a cardiologist. I write everything down so I don't forget, but I still miss some appointments because I can't keep track. I've been diagnosed with a bunch of stuff that doesn't really make sense. They thought I might have Lyme disease. They say I might have something called POTS, or chronic fatigue, or fibromyalgia. I have this big pillbox now, and a lot of the medications are experimental or for off-label use, so we pay a lot of that money out of pocket. I take two antidepressants, vitamin D, and a whole bunch of other stuff. My husband has to keep track of the schedule because it's too much for me. I cashed in a favor with an old high school friend, and he got me an appointment to see a neuro-infectious disease doctor at Northwestern who works on long COVID. He assessed me and gave me a cognitive test, which I failed. He said there are probably tens of thousands of people like me who are having long-lasting neurological effects from this, but it's going to be years before we really understand it. He said: "We're just beginning to learn. Physically, you look okay, so it's hard to know the right solution."

If nobody knows what's wrong, how do I get better? My vitals are usually normal. My lung scans look fine. My blood work turns out to be okay.

It sounds crazy, right? Am I crazy? I definitely have that psychological battle in my head. I start to doubt everything. Maybe I've gotten lazy or something and it's all in my head. I'll force myself out of bed, but then I get in the shower and my heart rate spikes. The hot water turns my hands purple. I get so dizzy I have to sit down.

I'm lucky to have this great support system, but I need help with everything. I can't really drive. It's exhausting for me just to be a passenger in a car. My husband and I moved out to the suburbs to be near my parents. He's the full-time worker, the full-time caretaker, and the full-time housekeeper. He's amazing, but it scares me. This isn't the marriage he entered into a year ago. He has a walkie-talkie and he comes to check in on me every hour, and meanwhile I'm like this helpless ten-year-old, just lying in bed and trying to keep my mind busy. I play some basic video games. I look online at different house décor. A doctor told me that doing arts and crafts is a good way to keep my hands active, so I sat in bed on Halloween decorating little paper pumpkins. A few months ago,

I was making million-dollar trades and traveling all over the country. Now I draw little scarecrows and tape them up on the wall. "Good work, Kaitlin! You're using your brain. You should be so proud."

It's guilt. It's anger and a lot of self-loathing. I have therapy once a week, and that helps. We talk a lot about acceptance. I'm trying to accept the fact that I have cousins in Wisconsin who are still posting on Facebook about how this virus doesn't actually exist. I'm trying to accept that I might not get better. I lie in bed and tell myself not to think about anything but the present moment. I try to let go of my expectations, but it feels like surrender.

Chapter 24

"Do people understand what's happening here?"

Bruce MacGillis, on the excruciating wait
for a vaccine inside a nursing home
Mentor, Ohio—December 2020

I'm happy they put us at the top of the vaccine priority list,
but I doubt it's going to make much of a difference in here.
Can I get the shot today? Will I have immunity by tomorrow? 'Cause that's the kind of timeline we need to be thinking
about in this nursing home. More than half of the residents in

this place are COVID-positive. I'm one of about eighty residents, and thirty got sick *this week*.

The first thing I do when I wake up is look down the hallway for the big plastic sheet. That's what they use to block off the COVID area. They sectioned off a whole wing of our building a few days before Thanksgiving. Then they blocked another hallway earlier this week. That plastic sheet keeps moving closer. I'm trying not to panic, but where am I supposed to go? It's not like I can jump up and make a run for it. I'm in a wheelchair. I haven't been outside for months. I'm trapped, just like everybody else in this place. We're at the mercy of this virus. We sit here and we wait.

That's been the story of the last nine months. It's boredom and then dread. They stopped allowing visitors in March, so we lost that contact with the outside world. Then it was no more group meals in the cafeteria—just eat everything alone in your room. No more trips to physical therapy. No more access to the lounge or computer area. My world keeps on getting smaller. I have my little room. I have my old nine-inch TV. I play Sudoku and watch Turner Classic Movies and stare out the window at the woods. I check the latest COVID numbers every few hours on my phone and try not to have a heart attack. It's up to 300,000 deaths in this country, plus we're adding another 200,000 new cases every damn day. The administrators in the nursing home keep telling me not to panic. Are you freaking kidding? We better panic. All of Ohio is out of control. We managed to keep the virus out of here for a while, but with the numbers this high, it was just a matter of time. COVID eats places like this alive. I keep reading about how more than 100,000 people have died in nursing homes just like this, and I don't want to be one of them. I make it from one sunrise to the next. I keep breathing. That's it. That's the whole goal.

I saw my doctor a few weeks ago, before this whole outbreak started. I'm only sixty-four, but I have a lot of issues because of my accidents. I used to drive a newspaper truck, and one night I was sitting at a red light with the Sunday paper when I got plowed into by a drunk driver. I have balance issues, immune suppression, lung clots, weight problems, high blood pressure. I've been here twenty-three months trying to get myself better and get home. My doctor told me I have at least nine of the markers that are really bad for COVID. I asked him: "Are there vitamins or supplements I can take? Exercises? What should I do to protect myself?" He said: "Just make sure you don't get it."

Don't get it. Don't get it. Don't get it. That's what my brain does all day.

I don't let anybody come near me anymore. I have a little dresser by my door where the nurse aides put my medication and my meals. I stay away. I wait and then I put on my gloves to pick up the food, and that's how I eat. Otherwise, I keep my door closed, and I stuff some towels under it to block any air from getting through. I've got a window in here that I keep open no matter what for airflow, even when it's like fifteen degrees. There are a few nurses in the morning that I trust, so that's when I ask for help and take my shower, but otherwise I'm hunkered down. A few months ago, an aide from a temp agency tried to come in to check my vitals. We didn't have any cases at that point, and I thought some people here were starting to get a little casual about it. Her mask was down by her chin. She wasn't wearing any gloves. I'd seen her with some of the other aides talking in the break room, and they weren't social-distancing up to my standards. I told her: "Stay out of my room." She said she needed to check my blood pressure. She said: "I'm just trying to do my job. Why are you making this into such a big deal?" I'm a hard-ass about this stuff, and I'm not even a little bit sorry. I can't afford to take chances. I called our administrator to log a complaint, and finally she turned around and left. I told the administrator: "I've studied all the protocols. I know you can't leave me in the dark on this. If you get a positive case, you're required to let me know."

Two staff members came to my door the Friday before Thanksgiving. They told me one resident had tested positive. They didn't say how the virus got in. We have staff shortages all the time, and new temps are always coming through. They live in the community. They go work in one nursing home on Monday, and then on Tuesday they take a shift at another nursing home. They have kids. Their kids go to school. This last month, one in every fifty people in Lake County tested positive. The virus could have come in from anywhere. I wrote down the time and the date in a notebook I keep next to my bed. "Okay. That's one. Here we go."

The next day we had another case. Sunday it was two more. Monday we had eight positives. Then a bunch of the staff started to call out sick. I keep on writing it all down in my notebook, but it's getting hard to keep up. Last time I counted our cases, we had forty-five residents and twenty-one staff. That's all in the last two weeks.

We need help. Shouldn't that be obvious? The residents are scared. We can hear the beeps, the patient alarms, the ambulance sirens. I look out my window and I can see people leaving here in body bags. We have a good core of people who work here, but a lot of our management is out sick. They keep

bringing in new people to fill the gaps, and I know they're trying hard, but this virus keeps spreading. I'm starting to get desperate. I called the county ombudsman to alert him, and he's
supposed to be our advocate, but he said a lot of nursing homes
are going through outbreaks right now. He said thirty new
cases in a week isn't really considered all that extreme. "Okay.
Wonderful. Thank you for that." I called the county, but that
didn't go anywhere. I called the CDC hotline. I called a few
local Catholic priests. I called a number I found on Facebook
for Dr. Fauci, but that was just another message machine. The
Ohio Department of Health finally got back to me after four
or five days. They took down a report and said they would
notify a supervisor, but there are probably two hundred of
those reports waiting around in the same file. I haven't heard
back. I'll probably never hear back. It seems like everyone has
just surrendered.

I leave messages until my phone runs out of batteries. I
charge it up and try again. The other day I spent six hours making calls. I got so fed up that I dialed 911. It's a recorded line, so
at least that way I'm leaving behind some breadcrumbs. At least
they'll have something to go back to if they come in here one
day and find thirty or forty residents dead. I explained what
was happening to the dispatcher, but she didn't seem to get it.
She kept asking if I was in physical pain. She asked if I wanted
an ambulance. I said: "No, no. It's bigger than that. We're sitting ducks. Don't you get it? We all need to be rescued."

She said: "Sir, what's your emergency? I'm not hearing an
emergency."

I don't know what I expected her to do. She told me I
wasn't being rational, and maybe she's right. But why is there
never any acknowledgment? Why isn't there any urgency? At
least ten people are probably going to die in here based on our

numbers, and the way things are going it might be a lot more. What qualifies as an emergency? Why isn't anyone alarmed? It feels like I'm on the *Titanic,* and we're sinking, and I'm trying to make contact with the outside world using two soup cans and a string. "Hello? Hello? Can anybody hear me? Is *anybody* going to do *anything?*"

I get this sense sometimes that people are thinking: "Oh, it's just another nursing home. It's not a real tragedy. They were already at the end of their road." And for a lot of people in here, that's true. This is their last stop. But they're still people. They're still alive. There's one lady in here, and she's probably ninety, and every day she steals cookies out of the cafeteria and acts like she just baked them by herself. She puts on her lipstick and goes from room to room handing out her cookies. When she comes in, it doesn't matter if you're hungry. You better take a cookie. It's what keeps her going. She needs to give it to you. But now the cafeteria's closed, and she's lying alone in a dark room like everyone else with no one to talk to and nothing to do. She's one of the ones who just tested positive, and her room will probably be empty pretty soon. There's no human connection, no life, no hope. We're wilting away in here. Can you understand that? You start feeling like you've been forgotten. Where is everyone? Do people understand what's happening here? Do they care?

I'm coming to the realization that it's up to me to watch out for myself. The cavalry isn't coming. Nobody's rushing in to save us. It's like the whole building is burning down but nobody even notices the fire. I managed to dig up a plastic picnic blanket in one of the linen closets, and I'm going to tape it up over my door. I put up a sign that tells people not to come in no matter what. Total isolation might be my only chance at this point. I probably have to survive at least another month in order to get the vaccine, and I guess that's when we'll become a priority.

CLAUDIA HARVIE, in Arlington, Va., on
caring for a parent in a nursing home

For nine months, I woke up with the same fear: What if today the virus gets into his nursing home? I kept thinking about him getting sick, lying there, dying all alone. None of us were allowed to visit. There was nothing we could do. We hoped. That's it.

He was in a memory care unit in Dallas, a nice one, for whatever that matters. My sisters and I had cameras installed in his room so at least we could see him. I would get on my computer in the morning and check on him. I could see the whole room. I'd watch him getting out of bed, and I'd look at his face to see if he was grimacing, or if his skin was pale. I'd watch for a while and then do some work on my computer before clicking over to check on him again. I'd say "Hi Dad," and sometimes he'd look up at the television in his room, where he could see me. I don't know what he thought, because his dementia was pretty advanced at that point, but I wanted to be there and the camera was as close as I could get.

The nursing home had some cases early on, but the big outbreak didn't happen until October. They had fifteen or twenty residents come down sick within about a week. I wanted my dad to stay in his room and isolate, but he would get lonely. He liked to go sit in his wheelchair by the nursing station and listen to them talk. One day I was watching on the camera when his doctor came into the room. I said: "How's my dad doing? Is he still okay?"

The doctor looked up and I could tell something was wrong. He said: "I'm sorry. I was about to call. He just tested positive."

I watched him start coughing, and then eventually his oxygen dropped. The nursing home transferred him to a hospital. We still weren't allowed to visit, but they had a video setup, and the nurses rolled his bed right up to the camera so we could see. He was in the hospital for more than a week, and I never stopped watching. I slept with my laptop on my stomach, and I'd wake up every fifteen minutes to watch the rise and fall of his chest under the sheets. I was scared to turn it off. After a few days, his breathing became erratic. It was these short, rapid breaths. His face got rigid and tight. I would watch for a while and then call the hospital. "Can you check him? I don't know if he's breathing."

He didn't open his eyes for the last four days. The only things he responded to were music and touch. My mother would ask the nurses: "Can you hold his hand?" Touch is a human need, but there's a very fine line when you're trying to press a nurse to risk her own life by holding an old man's hand when he has COVID. The nurses always wore gloves, and most of them were good about it. One of the nurses—she was amazing. She would caress his hair and his face. One night, she was holding his hand, and he brought it up to his mouth and gave it a little kiss. It was the last moment he seemed alive.

He was slowly slipping away from oxygen deprivation, but his heart kept beating. We got our whole family together over the video screen and held a vigil. Sometimes there were fifteen of us on there from all over the world, talking about how much we loved him. I'd like to believe he was listening to us, but I don't really know. You create this narrative in your head to cope: He can hear us. He knows we are here. He understands how much he's loved. But it could have been something completely different. He could have felt distressed and confused by the video, or alone, or disoriented by all those drugs. Sometimes the nurses would pull back his sheets to change him, and his hands would start shaking uncontrollably. It was all so distressing. I won't ever know what he experienced, but my faith—my faith as a daughter who loved him—is that he felt loved and cared for. He didn't feel alone.

His breathing got more labored. You could see his head jerking because he was trying to get air. The nurses gave him more medication, and he seemed peaceful. I drifted off to sleep and woke a few minutes later when I heard his nurse walk in the room. I looked at his chest, and it wasn't moving. I asked the nurse if he was okay, but I already knew the answer.

The nurse said she could give us some time alone with his body to grieve. She walked out of the room, and we stayed there on the video, crying and telling stories. The nurse came back in after about an hour. She said she was going to clean him up and take away his body. She shut down the camera, and the screen went dark.

Chapter 25

"I needed something good to happen"

Dr. Valerie Briones-Pryor, on ending a year
of grief with a moment of hope
Louisville, Ky.—December 2020

I was one of the first people to get vaccinated in Kentucky.
The whole thing was surreal. It happened in a small auditorium, and the governor gave a speech. I walked onstage and
pulled up my sleeve with the cameras rolling. A few people
clapped when the needle went in. Some of us were crying. It
felt like this amazing victory celebration, and then I went back
over to the hospital to check on my patients.

One coded on me that morning. Oxygen deprivation. He
was my twenty-seventh COVID death. Then I had another
guy who's been with me fourteen days, and I thought he was

finally getting better, but COVID schools me all the time. Suddenly, he couldn't breathe while he was doing his physical therapy, and we had to rush him onto one hundred percent oxygen to get him stabilized. Then I checked on a patient who can't keep anything down. She's young, and her breathing is fine, and she should have gone home by now, but that's not how this virus works. Lately, it feels like my batting average isn't very good. I sent one patient home, but I had four more going in the other direction and getting worse, so I transferred them to the ICU.

So, yeah. That's how it's gone lately. I guess the best part of my day was a little arm soreness.

I'm desperate for this to be over. That's why I'm so thankful for this vaccine. It's safe. It's effective. It's miraculous. We got a vaccine for this virus within a year of it showing up for the very first time. It defies all expectations and all medical history, so I wish I could sit here and say: "Okay! That's it! We're done!" But even with the vaccine, the reality in our hospital hasn't changed. The case numbers are still high, and we've got months more to go. I've been in charge of our COVID unit since we opened March 17. It's wave after wave. Treat COVID. Study COVID. Worry about COVID. That's all I do. I shower at work, change clothes at work, and then wrap my son in a blanket before I give him a hug so I don't bring COVID home.

We started our COVID unit with fifteen beds, and that was enough for a while. Then we moved to another unit that had twenty-two beds, and now we need both units. I usually get patients in that eight-to-fourteen-day window after they've contracted the virus, when they could go either way. They come through the emergency room and get admitted to me. Most of them are here for a while—could be a week or ten days. We use steroids and sometimes an anti-viral, but a lot of hospital medicine comes down to observation. You're examining a patient and looking for clues. What are their breathing patterns? How do they look when they're eating? Has their color started to change? Our nurses are so dedicated, and they monitor these patients twenty-four hours a day. They're not allowed to have any visitors, so I try to sit in their rooms when I can. I get to know them. I do a lot of watchful waiting.

I started thinking back over some of my patients while I was getting the vaccine—my list of twenty-seven. In a normal year, I might lose a total of four or five patients. This has been a lot to handle. A few weeks ago, I had nine deaths in nine days. It's been a lot of older people, and some had made the decision that they didn't want to be on a ventilator. We had an older Hispanic gentleman, and he didn't speak English, so I had to communicate to him using an iPad as our interpreter. He got so scared at the end that he couldn't be alone. He didn't want the nurses to leave his room. I had a thirty-three-year-old who kept getting worse for a while, and then I had to tell him he was going to the ICU. He tried to negotiate with me. He was wearing this high-flow oxygen mask, and he was crying all over it. He said: "Please, give me one more day. I'm begging you. Don't give up on me. I know I can get better."

One of my first deaths was a Catholic priest. He'd come from a nursing home that had an outbreak, and four of those patients died. It was my mission to make sure he got anointed before he passed. I called up my supervisor and said, "I know we're not allowing visitors, but I really need this favor." Luckily, we were able to get a priest who volunteered to come, and we got him all dressed up in PPE, and he gave last rites. It was beautiful. I'm Catholic, and after everybody left I sat there and held his hand. I'm a cantor at my church. I sang to him and told him it was okay to go.

I try my best to be connected. It's an honor to be with someone in those last moments, but it should be a loved one. It shouldn't be me.

This pandemic has taken me through all the stages of grief. There was that initial phase of denial, and some people in this country got stuck there and never moved on. Then it was anger, and I definitely had that. I was mad at some of our politicians. I was mad at my neighbors because they were having people over at their house. Then I was bargaining—maybe if we lock down the economy or do this or that, it won't be so bad. Then depression. Then acceptance.

But the problem with acceptance when you're in the middle of a pandemic is you start to get numb. I've gotten numb to where I almost couldn't feel anything. For a while, I could just come in to work and do my job and put that smile on my face and deal with it, but eventually it starts to get to you. I worked twenty-one days in a row at one point. You get beaten down. It's never-ending. You discharge four people and then five more cases come in a few hours later. What's the point of fighting this virus if the virus always seems to win? At some point, you almost have to depersonalize the patients you're treating to get through, and that doesn't feel right, either. You start thinking: I'm done. I'm on empty. I don't know how much longer I can do this.

A few weeks ago, I started having heart palpitations. It's something I see in my patients, because COVID can impact your heart, so the first thing I thought was: Oh no. What if I finally got it? My son overheard me talking about that possibility to my husband, and he started to cry. He said, "I don't want you to die." He's seven, but he's smart, and he pays attention to everything. He sees how we've changed our lives to avoid this virus, so now he thinks that way.

I got tested and it was negative. I went to see my cardiolo-gist. My blood pressure was up. My doctor said it was probably stress built up over all these months. We were up to having, like, thirty-five COVID patients between the two units at that point. My husband was getting worried about me. One of my medical partners said: "You can't keep going like this. It's too much." We decided I'd hand over the second COVID unit, so now I'm just in charge of one, and that's enough. I'm exhausted.

I needed something good to happen—something to pull me out. As soon as I heard we were getting a shipment of the vac-cine, I put my whole heart into that. I've been waiting for this a long time. I mean—I'm tearing up now just thinking about it. It's great to know I'll have protection against this virus, but it's more than that. It's a profound relief. I can finally see a way out of this, even if we aren't there yet. It's a beginning. It's a reason to hope.

Chapter 26

"Nobody told us what to be ready for"

Roger Desjarlais, county manager, on the
challenges of rolling out a vaccine
Fort Myers, Fla.—December 2020

We're trying to get this vaccine into people's arms as fast as we can, but everyone seems to have a better solution than the ones we're using. I'm not very popular right now. I've been called incompetent more times in the last month than I have in my whole career.

Could our vaccinations have been more efficient from the get-go? Absolutely. I could spend all day trying to pass around blame and dodge responsibility, but I've been in public service for forty years, and part of being in a leadership position is knowing how to stand up and take some hits. We're trying to get better. We're figuring it out as we go.

I'll be the first to admit that I didn't expect for us to be managing this rollout at the county level. For whatever reason, I made the assumption back in the fall that when vaccines became available, it would be handled by some combination of federal and state government. Why I didn't think beyond that, I can't tell you. I'm feeling a little stupid right now. Each state was left to figure this out. The state handed the operations piece on to the county. That's not what I anticipated. I've gotten a lot of emails from people saying: "You've known for nine months that this was coming, and you should have had a plan in place." But the truth is, nobody told us what to be ready for. I had no idea this would be our responsibility. I'm sorry, but I just didn't.

There had been rumors for weeks about vials of vaccine starting to show up in Florida, but we didn't know where, or which one, or if it would be one dose or two, or who would be eligible to get it. I got a phone call the Saturday after Christmas from our local health department asking for my help, and that was the first time I got involved. The state had given us 12,000 doses of the Moderna vaccine with instructions to administer it as fast as possible to health-care workers and anyone over age sixty-five. There wasn't a whole lot more information. We asked the higher-ups: "How much more vaccine are we getting? When should we expect to have more?"

Nobody knew. We've been given a lot of responsibility when it comes to giving out this vaccine, but not much control. It was basically: "You know as much as we do. Go get shots in arms."

The best and fastest way we could come up with was to

have people line up—first come, first served. You get them in the same place and then it's "next, next, next, next." We have 240,000 seniors in this county. Cases are spiking, and we've got this new strain to worry about. It's a race. We have to be fast. I'm lucky that I work for a great county commission, and we have a big team. The sheriff got involved, the department of public safety, emergency management. We opened up community sites, put out a press release and got everything together within about a day.

We expected to have a big demand, but I don't think anyone could have anticipated just how much. People started lining up ten, fifteen hours before we opened. They camped out overnight, even though we asked them not to. So now you've got seniors coming with blankets and medications, and they're sleeping in chairs in a parking lot. That's not what anyone really wants. It got a little chilly overnight. The criticism was extreme, and I understand it. I recognize that's a hardship for some people. We had one gentleman fall down while he was waiting in line, and the ambulance came, and he got his vaccine in the back of the ambulance because he didn't want to miss out. We had snowbirds, tourists, people caravanning from other parts of the state. If they're over sixty-five, the federal and state rules prohibit us from turning them away. So now you're dealing with hard feelings about that. You're dealing with traffic issues, because you have two thousand people showing up at a rec center. There's line-cutting. There's anxiety. We don't have a mask mandate in Lee County, so we can't force people to wear masks, and that's another issue in line. We asked them to maintain social distance, but a lot of people don't. A few people complained it was going to be a super-spreader event, and I'm not so sure about that, but you certainly don't want to be doing more harm than good. The police brought a mobile watchtower, and twelve nurses administered vaccines. You had seniors who are breaking down in tears because they're so grateful and relieved that they can finally start seeing their grandkids again. We were getting through two hundred vaccinations an hour, so that's a very happy crowd. But then you had other people who waited for hours and couldn't get in, and they're getting panicky. They're cursing. They're trying to call in favors. They're pleading with you even though there's no vaccine left on-site.

So it's this tremendously emotional event, and afterward you look at the numbers, and the math isn't very forgiving. That first week across all the sites, we administered a total of about five thousand doses. That's a tiny percentage of our seniors. Then we still have to do the second dose. Then it's the rest of our population.

We could trudge along for months and months, and that's not an option. We have to do more. Last week, we decided to construct a centralized site for vaccinations out at the airport to build our capacity. We hired people to manage traffic flow. We got a big air-conditioned tent from New York. We trained more paramedics to give out the vaccine. We set up a phone reservation system through a vendor so people could call in and get a guaranteed slot.

We had five thousand slots for appointments spread out over the week, because that's all we could accommodate until the state gives us more vaccine. The 1-800 number went live Monday at noon. Within the first four minutes, people started to get a busy signal. Thirty minutes in, we'd hit 1.5 million calls. These residents are smart and they're determined. We looked at the phone records and one woman had called something like 236 times. People would enlist fifteen relatives from around the country to keep calling on their behalf. We had situations where ten seniors would get together in a room, and if one got through to somebody, they'd make their appointment and then say: "Hold on. The guy next to me needs an appointment, too." The system was flooded. It started to fail. People were getting odd-sounding beeps and error messages. Apparently, some calls got intercepted by scammers trying to sell vitamins. Other people were put in a queue to get a reservation time, and they waited on hold for three or four hours and then suddenly got cut off.

I hear about all of it. It's emails and more emails. Hell, I'd be mad, too. I know the call center can fix the glitches. That part we'll sort out. But with a lot of these emails lately, they're asking me questions I still can't answer: "How many more doses will you give this month?" "I got my first shot. What about my second?" Look, if I knew right now that we had 50,000 doses to give every week, we could set it up on a schedule and really manage this thing. We could scale up. We could do it fast. We could take more control. But I don't know when we're getting more vaccines. I can't create a long-term schedule. I don't know how many doses will be in the next batch. I don't know what to tell people about their second shot, because we're still waiting on that information from above.

The message we keep getting from the state and federal level is it all depends on the supply chain. They say we need to be prepared to roll with the punches as they keep coming.

There's only one way to satisfy everyone. We got 5,000 more seniors vaccinated last week, and they're happy. That means we have about 230,000 left to go.

MARLENE ROEHM, resident, Fort Myers, Fla.

We have three phone lines in the house, and as soon as the call center opened, we both started dialing like crazy. I'd already told all our friends: "Don't call us Monday afternoon. We need the phone lines to be free." I'd do anything to get my hands on this vaccine. I was acting like my life depended on it.

We're old, and we're pretty vulnerable. My husband has some health issues and he has to take blood thinners. I know the whole country probably feels this way, but sometimes I don't think I can live like this for one day longer. I've spent a full year hunkered down like a hermit, spraying down my mail with seventy percent alcohol. It's fear. It's anxiety. I mean, what happened to my life? I really, really need this to be over.

I was on my cell phone. My husband, Bill, was going back and forth between the landline and the computer. It was just dial, dial, dial, dial. We would look over at each other after each failed attempt and just shake our heads. I looked at the phone record later, and I'd called the hotline 137 times in the first half hour. Bill called 42 times from his cell and 40 more from the landline. We kept getting a busy signal or one of those recorded messages saying all circuits are busy. At some point, panic kind of started to set in. *Please*—pick up. Pick up! I want to see friends again. I want to be able to visit my daughter.

After a while, Bill clapped his hands and motioned me over. He said: "I got somebody!" The guy on the other line had a thick accent, and Bill has a hard time hearing, so he gave the phone to me. I talked to the guy for about four minutes, but he wasn't making any sense. He kept saying something about buying vitamins. He said: "You will be strong. You will be healthy!" It was very strange. I said: "No. No. I don't care about vitamins. We want the vaccine!" He said he didn't know what I was talking about. I hung up and dialed again and again.

Bill finally hit the jackpot on our 248th call. The lady on the other end was very nice. She asked for all our information and then made the appointments. She said: "Our phones have been totally jammed. You got very lucky."

We went out to the airport a few days later to get the shot. It was nothing. It was easy. I'm a retired pharmacist, and I worked in a hospital, so I don't mind needles. I didn't even notice when it went in.

They had us sit around for fifteen minutes in a holding area, just to make sure nobody had any complications. There were probably twenty of us who had just gotten the vaccine. We were spaced out in chairs and everyone was so joyful. We started talking to each other, laughing, telling stories about the pandemic. I hadn't met anybody new in a long time. I'd been scared of other people for a whole year, and it felt like a new beginning. I turned to the lady next to me and I introduced myself. I said, "Hi, I'm Marlene. Can you believe it? We made it!"

Chapter 27

"The longest and shortest year"

Stanley Plotkin, legendary virologist, on a
time of suffering and scientific progress
Doylestown, Pa.—January 2021

I've been so focused on helping to develop these vaccines that I barely thought about the mechanics of getting it myself until this month. How can our process be this complicated? I've been calling around now for the last several weeks. I could not find out where, or when, or how to receive a vaccine. I didn't get anywhere.

I'm eighty-eight years old. I'm in the priority group, and I qualify by all the Phase 1 recommendations. That should be enough. I don't want to jump ahead of anyone in the line. I don't want to call somebody up on the phone and say: "You know, I actually consulted on this vaccination process. I wrote the textbook on vaccines."

If I'm not able to get in through the normal channels, that means a lot of people like me are not getting in, and that's a big problem. We shouldn't be experiencing this level of chaos in a developed country at a time like this. I live in a suburb of Doylestown, which is the county seat here in Bucks County, so it's not like I'm way out in the hinterlands. I did research online and registered for all sorts of things with the state and county, and I never heard back. I called and called and couldn't find a way to get vaccinated within twenty miles of my house. My wife got frustrated on my behalf and started searching outside the area, and just by chance a few days ago she found a hospital in a different town that had openings left for the Moderna vaccine. I've barely left home to get the newspaper, but I was lucky to find this vaccine anywhere. I realize a big part of the problem is the lack of supply, but millions of people are being left on their own to navigate this disorganized mess. It's a free-for-all. What kind of system is that?

The whole experience of this pandemic has taken us through the emotional extremes during these last twelve months. It has been both the longest and shortest year of my life. There's my frustration with our poor national response, which has not been proportional to the threat of this virus. But I also try to focus my mind on the encouraging moments, and we've had many. Our vaccine development has been remarkable and entirely unprecedented in the history of mankind. There are many reasons for doom and gloom, but the science is sustaining.

Let's not exaggerate my own role in all of this. I don't have a laboratory anymore, so all I can do is consult. I've been offering my advice to the World Health Organization and the Gates Foundation. I serve on a board for Moderna and communicate with Oxford, Sanofi, and Inovio. I helped create the Coalition for Epidemic Preparedness Innovations, and it's supporting the development of about eight vaccines, but I can't take credit for what's come down the line. I've been cheering from the sidelines and shouting out my suggestions. I've been incredibly busy without ever leaving my house.

What I've been focused on lately is advocating within these organizations for a change in our vaccination strategy, so we get one dose to as many people as possible over the next few months. I've looked over the data, and it suggests that you have pretty good immune protection starting twelve days after the first dose. Obviously, you need a second dose to get long-term protection, but the immune system stays primed for the second dose for at least six months. I'm suggesting we wait on the second dose for up to twelve weeks until our supply improves. Otherwise, a lot more people are going to be calling around and striking out like I've been doing. It's a controversial approach, but an extraordinary disease requires extraordinary solutions.

Now, will anybody listen to me? I don't know. One of the advantages of being older and somewhat respected is that people are polite about soliciting my advice, even if they don't always take it. They value my experience. I cut my teeth on polio and anthrax in the 1950s. I developed the rubella vaccine that's now in standard use throughout the world, and I've worked on vaccines for rotavirus, rabies, Lyme disease, and cytomegalovirus. When I first started, we only had two ways to develop a vaccine, and now we have many more methods that show incredible promise. More than a hundred vaccines are being developed against this virus—all in record time. A year ago, we barely knew this virus existed, and now we've created a viable solution that has the capacity to end this pandemic. These vaccines will save millions and millions of lives.

Sometimes I force myself to stop and consider that, and it makes me physically dizzy. I think back to the Black Death, which killed something like thirty percent of the population, and something like that should not be forgotten. For most of my career, the process of creating a vaccine took a minimum of five years, and three out of every four vaccines never made it to the market because they didn't protect well enough or they caused too many reactions. We have gotten so much more advanced in our vaccinology. Science is cumulative. It builds steadily toward progress, and that's been my answer to despair during this last year. I can look back over my life and see a degree of advancement that's staggering.

We can say with justification that vaccines have changed the world, and that gives me hope that they can do so again. I had contracted three diseases by the time I was ten years old that are now prevented by vaccine: pertussis, pneumococcal pneumonia and severe influenza. At that time, only a handful of vaccines were given to children. Now at least sixteen are on the routine schedule. Parents can expect their children to grow up, and that's a relatively new thing. It shouldn't be taken for granted. But because people now have the great luxury of forgetting about these diseases, we are starting to run into all kinds of strange conspiracy theories about vaccines. Some people revert back to the Dark Ages of mysticism and pseudo-science. The White House had that guy [Scott] Atlas. I mean, my God! The minimizing, the skepticism about masks—you couldn't have made it up. Then there are people like Andrew Wakefield or Robert Kennedy, who have influence and use it to spout nonsense about vaccines, and that's dangerous.

If I were president, every child in the country would be required to take a course in statistics. Opinions don't count for all that much. Facts count, and we have a lot of data about the safety of vaccines. Nobody can say vaccines are one hundred percent safe. There's nothing that's one hundred percent safe. That's ridiculous. But there is a system in place, called the National Vaccine Injury Compensation Program, to financially compensate people for vaccine reactions, in which the evidence is presented to a panel that weighs the facts and then decides. It works out to be about one compensable reaction per every one million doses of any particular vaccine. So then you have to ask yourself: What is the risk of the disease at the moment? This virus is replicating. It is constantly making more of itself, and during that process there are always variations and mutations that occur. I don't think there is any doubt that the risk of this disease is a hell of a lot bigger. Any reasonable person should accept vaccines on the basis of logic. Unfortunately, logic does not predominate around the world.

Organisms, viruses—they are not going away. They will continue to cause epidemics. This virus will likely persist in the form of sporadic cases and occasional outbreaks no matter how hard we try. What we may be looking to is an influenza-like situation, with a virus that mutates over time, and we will have to tweak the vaccine sporadically to meet those changes. I hope I'm wrong, but I'm not optimistic this virus will entirely disappear. We live in a world where you can wake up in Kampala and go to bed in New York. The spread of viruses through human agency is more likely now than ever.

It's not an exaggeration to say our future depends on finding solutions. That's why I wanted to work in a laboratory ever since I was about fifteen. It is tedious and uncertain work, but it has aspects of an almost religious experience. You are that explorer in unknown territory. I think back to a Tennyson poem, "Ulysses," which I quoted long ago in my college yearbook: "To follow knowledge like a sinking star, beyond the utmost bounds of human thought."

Say all you want about the horrors of this last year, but we have reached those utmost bounds. We have accomplished more than I would have imagined possible, and it's been a global effort. It's astounding. It's thrilling. I was about to say it's miraculous, but that's not right. I don't believe in miracles. It might sound a little fancy, but I believe in science. I believe in our capacity to endure and overcome.

Postscript

Tony Sizemore waited a full year to hold a proper funeral for Birdie Shelton before deciding instead to have her ashes buried in a small family graveyard. "She would have wanted a big party, but we can't do that if everyone's still getting sick," he said. "It's another loss for Birdie. I'm still not right in the head about any of this. I won't be right in the head for a long time."

Sizemore was never tested for COVID-19, but three of Shelton's co-workers at Enterprise Rental Car in Indianapolis also died of the virus, along with more than eight thousand Indiana residents by the end of 2020.

. . .

Sal Hadwan and four other nurses in Detroit filed a lawsuit against the parent company of Sinai-Grace Hospital, alleging that several patients died unnecessarily during the first month of the pandemic because the hospital was overrun and understaffed. According to the lawsuit: "Dozens of patients perished due to the inability of the overloaded medical staff to get to them, monitor them and provide treatment, including those that were only discovered to be deceased after they had died and rigor mortis had set in."

The hospital has denied any wrongdoing. Hadwan, Mikaela Sakal and five other ER nurses who worked on the night shift left the hospital to take other nursing jobs.

. . .

BURNELL COTLON continued to allow his customers to buy groceries with store credit for several months, eventually lending out more than $45,000 and falling three months behind on his own mortgage. When people in New Orleans and beyond started to hear about his generosity, they returned it tenfold, raising more than $500,000 for his store in a series of online fund-raisers. Cotlon used the money to forgive his customers' debts and begin construction on a subsidized apartment building next to his store. He also gave out free school supplies and turned his store into a free vaccination site for the community.

. . .

CORY DEBURGHRAEVE performed hundreds of intubations during the pandemic and also managed to stay healthy. He was one of the first people in Illinois to be vaccinated in December.

. . .

GLORIA JACKSON continued to spend the year at home isolated from all physical contact. She joined an online group for women over seventy and celebrated Christmas with her family over Zoom.

. . .

MICHAEL FOWLER went on to pronounce more than 250 deaths from COVID-19 in Dougherty County, Georgia, during the first year of the pandemic. The county was hit by a second wave in January 2021 that resulted in almost forty more deaths, prompting the community to approve construction of a new morgue to accommodate more bodies. "I've been too busy to grieve," Fowler said. "I've stopped expecting it to slow down."

. . .

IAN HAYDON had no more side effects from his vaccine. Moderna's Phase 3 medical trial showed ninety-five percent effectiveness against COVID-19, and the vaccine was approved for use in the United States in December 2020.

.   .   .

After three more New York City EMTs committed suicide in the summer of 2020, ANTHONY ALMOJERA helped create a new program to provide paramedics with ten free EMDR sessions to treat post-traumatic stress. Almojera was one of the first to sign up for the psychotherapy treatments. "I've been in therapy for seventeen years, but this is my first time doing EMDR, and it helps," he said. "We still respond to COVID cases all the time, and it never gets easier, but we're taking time to process it. A bit of that heaviness is finally starting to go away."

.   .   .

JEFF GREGORICH defied the governor's mandate and chose not to open his school district in Winkelman, Arizona, for in-person learning. The school buildings remained closed for the rest of 2020. Several other districts across the state followed Gregorich's example, and his peers at the Arizona School Administration voted to give him the "Bastion of Courage Award" in September. When the Winkelman school district reopened for limited in-person learning early in 2021, a new portrait of former teacher Kimberly Byrd hung at the entrance to the elementary school.

.   .   .

TUSDAE BARR found hourly work in fast food and saved up more than $3,000 for a new apartment, but she still couldn't find a landlord willing to rent to someone with her history of evictions. She lived in weekly motels for several months before moving in with family members.

. . .

TONY GREEN lost one hundred pounds during his recovery from the virus and then moved away from Dallas, hoping to distance himself from the blame and hurt that caused a rift in his family after the gathering at his house. He relocated with his partner to Oklahoma City, but after a few months he began to wonder if he had moved far enough. "This country's response to the virus, our political polarization—there's nothing left to stop our demise," he said. "I will put America in my rearview mirror sometime in 2021. The last freedom I intend to indulge in here is the freedom to leave."

. . .

BRUCE MACGILLIS barricaded himself inside his room for twenty-three straight days in December as a wave of infections continued to spread through his nursing home in Ohio. By Christmas Day, eighty-seven of the eighty-nine residents in the facility had tested positive for the virus, but MacGillis remained healthy. He locked his room, stuffed towels underneath the doorway, wrapped himself in a blanket, and kept his window open despite subfreezing weather in order to maximize airflow. He refused to let anyone inside until December 28, when a nurse came in to administer his first dose of vaccine.

Acknowledgments

I'm indebted to the dozens of people who willingly set down their lives to speak with me for hours upon hours without expecting anything in return other than a chance to be heard. I interviewed more than two hundred people about their experiences with COVID-19. Only some of their stories appear in this book, but I learned from them all. In a time of divisiveness and disinformation, they answered a cold call from a stranger and gave freely of their time and their trust. This book is a result of their courage and generosity.

Many of these stories first appeared in *The Washington Post,* where I've been lucky to spend my career. I'm grateful to Marty Baron, Cameron Barr, Tracy Grant, Steven Ginsberg, Tim Curran, Bronwen Latimer, MaryAnne Golon, Greg Manifold, Andy Braford, Gilbert Dunkley, Jill Grisco, Maria Sanchez Diez, Stephanie McCrummen, Hannah Dreier, Chico Harlan, and most of all David Finkel, who helped to conceive and improve so much of this work. Warren Saslow, Rebecca Barry, Craig Saslow, and Alec Saslow read pieces along the way and offered meaningful encouragement. I'm thankful to my agent, Esther Newberg, for believing in this book and making it possible; and to Bill Thomas and the rest of the excellent team at Doubleday for so consistently supporting and enhancing my work.

My own story of the pandemic, like everything else about my life, was made better and happier in all ways because of Sienna, Chloe, Ari, and especially Rachel.

Illustration Credits

About the Author

Eli Saslow is a staff writer for *The Washington Post,* an author of three books, and a screenwriter. He's the winner of the 2014 Pulitzer Prize for Explanatory Reporting and a three-time finalist for the Pulitzer Prize in Feature Writing. His 2018 book, *Rising Out of Hatred,* won the Dayton Literary Peace Prize for nonfiction. He lives in Portland, Oregon, with his wife and three children.